Industrial Engineering Projects

Industrial Engineering Projects

Practice and Procedures for Capital Projects in
the Engineering, Manufacturing and Process
Industries

by

The Joint Development Board

E & FN SPON
An Imprint of Chapman & Hall

London · Weinheim · New York · Tokyo · Melbourne · Madras

**Published by E & F N Spon, an imprint of
Chapman & Hall, 2–6 Boundary Row, London SE1 8HN, UK**

Chapman & Hall 2– 6 Boundary Row, London SE1 8HN, UK

Chapman & Hall, GmbH, Pappelallee 3, 69469, Weinheim, Germany

Chapman & Hall USA, 115 Fifth Avenue, New York, NY 10003, USA

Chapman & Hall Japan, ITP-Japan Kyowa Building, 3F, 2-2-1 Hirakawacho, Chiyoda-ku, Tokyo 102, Japan

Chapman & Hall Australia, 102 Dodds Street, South Melbourne, Victoria 3205, Australia

Chapman & Hall India, R. Seshadri, 32 Second Main Road, CIT East, Madras 600 035, India

First edition 1997
© 1997 E & FN SPON

Typeset in 10½ on 12pt Times
by On Screen, West Hanney, Oxfordshire

Printed in Great Britain by Cambridge University Press, Cambridge

ISBN 0 419 22510 2

A catalogue record for this book is available from the British Library

LCCP no: 97–66020

 Printed on permanent acid-free text paper, manufactured in accordance with ANSI/NISO Z39.48-1992 (Permanence of Paper)

Contents

Preface

The Joint Development Board (JDB), which is sponsored by The Royal Institution of Chartered Surveyors and the Association of Cost Engineers, is charged with raising the profile of project and commercial controls in the Engineering Industry.

The JDB, which has previously published the *Standard Method of Measurement of Industrial Engineering Construction*, noted that there was no single source which imparted to the reader a clear, basic understanding of the manner in which industrial engineering projects were managed from feasibility through to commissioning and operation. The JDB also noted that a smaller workload and reducing margin had put all sides of the engineering industry under increasing pressure to improve the efficiency of their operations and the quality of its products.

Efficiency and quality are not restricted to design and construction activities. For a project to be a commercial success the project must be managed and the twin parameters of cost and time must be effectively controlled, by systems and procedures which are themselves subject to continuous improvement.

It was the recognition of the need to increase the efficiency and quality of project controls that prompted the JDB to produce this book, which aims to bring together the knowledge, skills and day-to-day practice of the engineering construction industry in the management and control of capital projects.

The Members of the JDB (including co-opted Members) responsible for the preparation and production of this book are:

A.E. Jackson, FRICS, MACostE
    (Chairman)
                      AMEC Process and Energy Ltd
D.R.D. Ainsley, FRICS   A.L. Currie and Brown
J.H. Blenkhorn, ARICS, MACostE,
    FInstPet   Franklin and Andrews
K.R. Cookson, FRICS   Franklin and Andrews
M.G. Leese, FRICS, FACostE,
    ACIArb   Bahrain Petroleum Company
M. Mitchell, ARICS   British Gas plc
P.J. McBrien, FCII, FIRM
D.F. Parkinson, FRICS, Hon. FACostE   Davis, Langdon and Everest
V. Thompson, FACostE
R.B. Watson, FRICS, MACostE   Engineering Cost Management
R.A. Webber, MACostE   Jacobs Engineering
B.G. Wheeler, FRICS, MACostE

On behalf of the JDB we wish to acknowledge the following for their invaluable help and specialist knowledge:

I. McCallum, MACostE, MInstCES	A. L. Currie and Brown
G. Davies, MIQA, CEng	AMEC Process and Energy Ltd
R.R. Genillard, DMS, MIMgt, MCIPS	AMEC Process and Energy Ltd
J. Roberts, BSc, CEng, MIMechE	Independent consultant

together with

Laporte Engineering Services for the use of certain forms

and

AMEC plc for the provision of photographs

We would like to express our thanks on behalf of the Councils of the two sponsoring bodies to all those who have contributed so much of their time and effort in the production of this first edition and for the support of their companies.

Alec Ray
President
The Association of Cost
Engineers

Jeremy Bayliss
President
The Royal Institution
of Chartered Surveyors

Foreword

Although the engineering construction industry is not necessarily in the public eye to the same extent as the building or civil engineering industries, it is nevertheless responsible for providing virtually everything demanded by a modern economy, from power generation and exploration and exploitation of raw materials to the manufacture of goods of all types.

Economic circumstances have made it necessary to review the manner in which all construction is undertaken in order to improve performance. The Latham Report into the building industry and the CRINE initiative in the offshore oil and gas industry recommend among other things a greater degree of cooperation between Owner and Contractor, integration of management teams, reduction in risk and standardization of documentation and procedures.

This book seeks to provide a greater understanding of the activities, systems and techniques used by members of the engineering construction design and construction teams, which is an essential start to achieving the changes envisaged.

It is recommended to those working in the engineering construction sector who are certain to find that it contributes to a greater general understanding of the industry. The broad spread of information also makes it suitable for students and for those intending to take up national vocational qualifications (NVQs).

Sir John Fairclough FEng
Chairman of the Engineering
Council 1990–1995

Introduction

1

The stripper reboiler at
Hydrofiner plant
Grangemouth, Scotland.

Plant Owner B. P. Oil Ltd.

Engineering construction is a loose expression covering the design and construction of capital projects for a broad spread of manufacturing and energy related industries including:

- gas and oil exploration and production;
- chemical and process;
- refining;
- nuclear;
- power generation;
- steel production;
- pharmaceuticals;
- oxygen and other gases;
- food and drink;
- water and sewage treatment.

Some of these industries are indigenous and have developed their practices alongside the construction sectors of other local industries, by adopting and adapting the same basic philosophies of administration and control, while others originated in other parts of the world.

Current practice has therefore been influenced by the nature of the projects typically undertaken and the traditional practices in the area of the world where the industry has its roots. In addition, owners and contractors in the above sectors are often among the world's largest companies, many of which have made major investments both in terms of experience and development costs in their particular method of doing business.

In spite of these historical beginnings many of the engineering sectors manage capital projects in a similar manner due to a number of common factors. These include:

- degree of owner involvement;
- remoteness of location;
- complexity of the projects;
- coordination of high value projects, with large multidiscipline, multi-contractor teams;
- time constraints;
- lack of standard documentation;
- hazardous nature of the plants being constructed;
- degree of risk involved;
- training of the construction professions.

The various factors that influence the economics of the completed plant will have been considered in the feasibility study before the project gets the go ahead. The plant may be competing against facilities built some years previously which have their costs written down, and the product from the new plant must match these existing prices by the use of new technology to increase efficiency, by reducing operating and maintenance costs, and by ensuring the lowest capital cost for the plant. This book primarily deals with the management methods and procedures necessary to achieve this economic success.

The project flow chart, Figure 1.1, shows the stages which are typical of an engineering capital project from conception to completion.

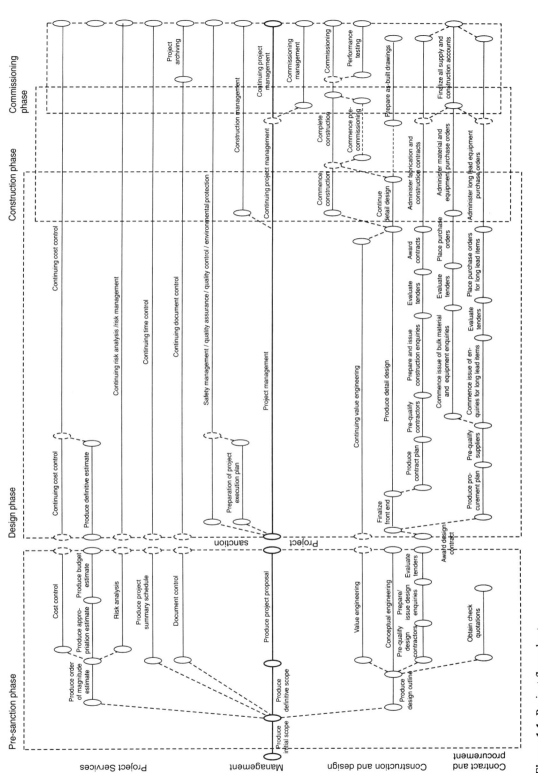

Figure 1.1 Project flow chart

This book does not cover all details of specific practices in the various sectors, but endeavours to provide an overview which has relevance to each sector.

1.1 OWNER INVOLVEMENT

Various factors affect the manner in which a prospective owner of a plant interacts with his design/construction team.

The typical owner involved in an engineering project tends to have greater technical experience and expertise than his opposite number in other construction industries in that he:

- may have similar plants elsewhere;
- will usually be the end user;
- may have design standards and in-house design capability;
- may hold a process licence;
- may have a project management capability including procurement, cost administration and contract specialists and will frequently maintain a supervision and coordination role;
- will certainly know how the plant can best be operated.

The typical owner of an engineering project will therefore be more involved in the day-to-day decisions and take a far greater degree of responsibility for the project than will most owners in the building and civil engineering industries. This involvement has a major effect throughout the whole organization of the work and on the contractual relationships between the various parties.

Generally, schemes take several years to implement and initially an owner must consider plant capacity, construction costs, production slots, feedstock, energy, transport and above all the available market and the price the product will command in the future. This will of course call for research as to what other organizations are constructing facilities, or planning to construct facilities, which would affect the product's market or price.

1.2 COMPLEXITY

Many engineering projects are by their very nature complex and include unique designs or processes. Owners demand the most modern technologies either to produce a new product or to ensure cost-effective production, etc. and frequently require the finished product in a short development period. This necessitates parallel working between the design and construction phases and the need for multi-contractor construction sites, with a consequent potential for delay, disruption and associated extra costs.

The design of a plant as a whole may incorporate designs provided by various specialist suppliers, who cannot design their component in isolation from the main design, but whose design is required by that main design. The resulting iterative procedure requires close control and expediting of information that flows between the various parties, in order that all the various designers, specialist suppliers and construction contractors work from the most up-to-date information at all times.

The time required to manufacture some items of sophisticated equipment may require that they are ordered ahead of any contract being tendered for their installation. Furthermore, bulk materials, e.g. pipe, valves, electrical and instrumentation cables, etc., may be purchased by the owner or the owner's managing contractor for reasons of compatibility or buying power. This gives rise to construction contracts where some or most of the materials are provided as 'free issue' to the contractor, which requires controls to be introduced into the manner in which those materials are purchased, issued and used.

On large projects, it is unlikely that any one main contractor will have all the necessary skills and resources needed to fabricate and construct the entire project. The result is the need to coordinate site access, availability of design and the free issue of materials and equipment, with the parallel or sequential working of a number of major construction contractors, subcontractors and equipment manufacturers.

Following construction, the plant will be pre-commissioned and finally commissioned and test run to verify the plant's performance. Some of this work may be undertaken by the owner, some by the management contractor and some by the construction contractors, subcontractors or equipment suppliers.

The impact of design complexity, the coordination of the design, procurement and construction activities and parallel working, is a major management function which highlights the need for tight control in terms of cost, time and information.

1.3 MANAGEMENT

When approval is given for a major project to proceed, many people are employed for months or years and many millions of pounds (or dollars, dinars, etc.) are expended.

The success of such a project rarely depends on a single individual. However, one single inadequate person at a sufficient level of influence can bring failure, overspend or delay (often all three).

The very term 'project manager' leads to a misunderstanding of the role of managers, and indeed it encourages the belief that projects can be 'managed' by an individual, whereas they can only be run by a team. The project manager is the leader of the team and as such must have adequate knowledge of the engineering issues, safety regulations and the law as it relates to the project, together with lots of common sense, the skills of a diplomat, an ability to face unpleasant issues quickly and enough respect from the team to get maximum effort and cooperation from them at all times.

1.4 INFORMATION CONTROL AND REPORTING

Understanding the importance of appropriate reporting is essential if a project team is not to be buried by the mountain of information generated by a typical major project. Large projects are too complex to allow project managers to regularly do any of the detailed work themselves, but a project manager does require relevant information to be provided in a condensed form in order to maintain an overview of the project and its key activities. The project will therefore need to have a bespoke service to give intelligble information in report form, supported by further detailed levels of information available to be examined, adapted and acted upon as required.

Unless the distribution of the large amount of information and reports generated by a project is controlled by a system which ensures that each member receives only what is required, the project management team can be overwhelmed by sheer volume of paperwork containing information which is constantly changing and developing. Moreover, unless such information is coded and correctly archived, the possibility of its successful recovery is remote.

Information control is therefore a significant activity on engineering projects, not only to ensure that all the various project members are working from the most up-to-date documentation and data but also to ensure that drawings and other documents are available to the owner which accurately reflect the as-built condition, since once completed the plant is an ongoing operation requiring maintenance and possible future development.

1.5 TIME

Time on a project is frequently all important, given the need to meet market expectation and customer demand; and in some cases the earlier earning capacity will outweigh any increased cost of accelerating the design and construction process. Time on all projects becomes increasingly critical as more of it passes. When the project is in the feasibility stage, a project completion and commissioning date perhaps five years in the future may seem a long way ahead; but when approval to proceed is finally received a year can have elapsed, but the completion date rarely moves. If the project becomes the subject of environmental objections further delays can accrue.

Notwithstanding environmental issues, the development of the design is the first stage of the project to feel the increasing pressure exerted on it by the passage of time; but the proposed completion date may still appear a long way off.

Further pressure is exerted on the design team as the procurement activity starts to demand design criteria and performance requirements in order that tenders can be sought for the purchase of long lead items. There is now an increasing awareness that time is indeed passing and that the time left to completion and commissioning is a little tight.

When at last fabrication and construction commence, time may have the whole project team by the throat. The time required for the logical sequential progression through the work may no longer be available. Design may have overrun and procurement be delayed as a consequence, but the end date remains cast in stone. The consequences of this are not difficult to predict.

When the programme slips, the reasons are often various or not specific. The need to accelerate the programme usually requires further expenditure by the owner over and above the original contract which, not surprisingly, he may not be too keen to incur. There will be a temptation to delay unpalatable decisions in the hope that subsequent events will make up any previous delays to the programme.

The overall effect of a delay to an owner may be loss of earnings and possible loss of markets and the owner will have to balance his options in deciding whether to accept the delay or bear the cost of accelerating the work.

The importance of the planning engineer's function in ensuring that the project is completed on time cannot be overemphasized. The many activities needed to design the work, procure equipment and materials, award contracts, coordinate contractors and suppliers, construct and commission the plant and set it to work needs to be carefully sequenced and progress monitored, and remedial actions need to be taken immediately to overcome problems.

The planning engineer will be faced with the difficult task of producing an overall project programme showing what is required, of whom and when, and then ensuring that all other parties involved in the project work to schedules which are compatible with this master programme.

1.6 SAFETY, QUALITY AND ENVIRONMENTAL ISSUES

Safe construction and safe operation is the subject of numerous Acts of Parliament and quality assurance is now generally used throughout the whole of the engineering industry. Similarly, the growing environmental lobby has also made owners and contractors alike more aware of the need to operate in an environmentally acceptable manner.

While health and safety is particularly important to the engineering industries in view of the danger inherent in any major industrial plant, if the desired quality is not achieved maintenance costs, guarantee of production and environmental protection can all be adversely affected.

It is now often the case that an operating or manufacturing company will demand that any contractor or subcontractor employed on a project must have quality assurance and quality control (QA/QC) procedures consistent with the requirements of ISO 9000, and that they also have an environmental policy. If a company does not have such controls internally, then it is unlikely to be considered when tender lists are prepared.

Environmental considerations are crucial to the programme. If the project is the subject of environmental objections during its development and goes to public inquiry, this may take 18 months, and a judicial review will take a similar period, while planning consents or emission licences will take about six weeks each time the project goes to committee. If, at the end of this period, another site has to be selected and purchased, this process will start again, with calamitous effects on the programme.

1.7 ESTIMATING AND RISK

Many engineering projects are of high capital value, and are possibly carried out in remote locations; many use processes at the limits of current technology, and many are unique or sufficiently dissimilar to each other to make it difficult to use a standard, high-level, cost database.

As a result estimates are required to be produced to a far greater degree of detail than in other construction industries in order for them to be sufficiently flexible to meet the demands of each unique environment and to enable the project manager to identify, analyse and closely manage the significant risks involved through the project period.

Large projects have special features which make them different and, when things go wrong, more expensive. Reasons can vary but include the following:

- The management of major projects is often undertaken by bespoke teams of specialists brought together for that one project and disbanded, after completion, with subsequent loss of team relationships and experience of working together.
- Unique items of plant may be ordered months, sometimes years before installation. Any mistakes or delays in the provision of this plant cannot be resolved by a trip to the local builders' merchant to purchase additional quantities or a replacement.
- The requirement to employ large numbers of specialist fitters, welders, etc. can deplete the resources of the locality, necessitating the investigation of the various options for off-site fabrication, importation of labour, etc.
- Separate disciplines and specialist suppliers may be designing their contributions in isolation, hence the possibility of error or change increases unless good coordination is practised.
- Engineering projects are site-specific, bespoke products and sometimes incorporate untried technology.

This list may have the effect of putting most of the contractors and team members, however experienced, on a 'learning curve' at all stages of the project.

Although there will always be a drive to improve management and project controls, the significant risks associated with engineering projects cannot be totally eradicated. It is widely recognized in the engineering industry that risk estimating techniques are essential and powerful tools, particularly during the early phases of a project. They are therefore used to aid the understanding of the risks and to identify areas where management attention is most needed to mitigate their effects.

The risks mainly arise from the complexity of the plant, the remote nature of the typical site, the urgency with which the plant is required to be completed and the areas of new and untried technology, all of which will test the most experienced estimator.

Although the aim should always be to improve performance and use 'fit for the purpose' designs and specifications, the decision to reduce cost by reducing specification has to be taken against the background of a realistic estimate which includes a proper consideration and evaluation of the project risks. To realize late in a project that features that were part of the original brief but omitted to reduce cost, could have been accommodated within the original budget will not endear the estimator to colleagues or employer.

1.8 COST CONTROL AND REDUCTION

The reality of a project will mean that unless effective cost control is exercised costs will escalate and the efforts of the most experienced of estimators using the very best of information will be wasted.

Control is not the passive role of merely monitoring and accounting for cost, nor is it limited to the cost of material and construction. Cost control is an active role, which commences on day one of the project, with the control of management and design, and it continues through procurement of materials and construction of the plant, to completion and settlement of all accounts. Cost control is undertaken by cost engineers whose function is to ensure that the work is undertaken in the most cost-effective manner, by seeking economic solutions, monitoring expenditure, analysing performance, identifying problem areas and recommending preventive or remedial action. The position of cost engineer may be filled by one who is by profession a cost engineer or quantity surveyor; however the term quantity surveyor is not widely used in the engineering industries.

Company costs can be reduced by various means including elimination of waste, removal of unnecessary requirements or by increased efficiency. Costs can also be reduced by reducing capability or specification, but whether this constitutes a 'saving' may be a matter of opinion since the owner is receiving less for his money.

Waste to be eliminated includes poor performance, lack of coordination and change for its own sake, all of which causes disrupted working and delay. The use of non-standard items, be they items of equipment, contract documentation, procedures or certification requirements, rather than an 'off the shelf' fit for the purpose, again incurs unnecessary additional cost or risk.

The CRINE (Cost Reduction Initiative for the New Era) report highlights many of these features in connection with the offshore oil and gas industries and emphasizes the savings that it considers could be enjoyed if the traditional confrontational forms of contract were replaced by a cooperative relationship between contracting parties.

Time will tell whether the objectives of the report and the Latham Report into the UK construction industry can be translated into the savings

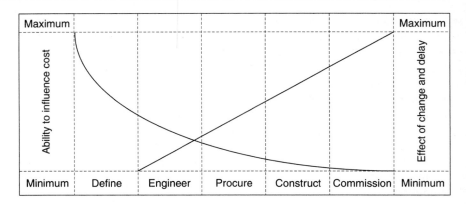

Figure 1.2 Potential for making savings and incurring additional costs during the life of a project

demanded and whether similar initiatives are equally valid in other construction environments.

It is sometimes necessary to challenge whether something perceived as essential to a project is in fact necessary. The removal of unnecessary content, or overspecification, from a project is a genuine saving and the use of the techniques of value engineering described in Chapter 4 challenges every aspect of a project and the components within that project. Value engineering is a technique which requires a systematic review of each element of the project and of every item of equipment within the project, questioning purpose and cost in order to identify savings.

Figure 1.2 shows that while the ability to make savings reduces as the project progresses, the effect of change and delay increases. This clearly demonstrates that to obtain best effect any cost saving reviews must be started sufficiently early. It also demonstrates that, during this same early period, management structures, controls and procedures must be established to improve and maintain performance and avoid the cumulative effects of change, delay, rework and disruption.

1.9 TERMINOLOGY

The lack of standardization within the industry is evident when looking at a list of apparently interchangeable terms which at best cause misunderstanding and at worst contractual error, dispute or other difficulty.

The need to fully understand the terminology used is obvious. A term apparently in common use may be given different definitions on different projects. For example, in this book the person for whom the project is being undertaken is referred to as 'the owner' while elsewhere the terms 'employer', 'client', or 'operator' are commonly in use.

Similarly the generic term 'change' is used to describe the following terms which are in general use to describe a change to the design, specification, means of construction, or timing of the work:

- engineer's instruction;
- variation order;
- trend notice.

Even greater difficulties relate to the term 'project manager', which is used in this book to describe the person in overall charge of the project. Some owners use the term project manager, while others refer to 'the engineer', as do some published standard conditions of contract.

To add to the confusion, some conditions of contract require both the owner and contractor to appoint project managers, while others may appoint a project manager to be in charge of the overall project, but use forms of contract which refer to the person ultimately responsible for a contract as being 'the engineer' or the 'owner representative'. Contractors and subcontractors may in turn appoint a project manager for their parts of the whole.

In considering alternatives in general use it is therefore important to consider not only the project as a whole but also the various contracts placed within that project. The function of a particular person or document should not be assumed without reference to a more precise definition peculiar to that project.

1.10 NON-STANDARD DOCUMENTATION

A major element having a significant effect on administration within the industry is the lack of standardization of contract documents, administrative practices, methods of measurement, etc.

This lack of standardization affects the efficiency of the industry in various ways:

- It reduces the possibility of amassing and exchanging data between projects and companies.
- It causes error in the understanding of requirements and requires tenderers to search through documentation to find potential contractual hazards and contractors to fall foul of clauses which they had misunderstood.
- It requires personnel to acquaint themselves with the systems and procedures for each project.

1.11 CONCLUSION

As in all endeavours teamwork and cooperation are vital ingredients to the success of engineering projects. An owner will rely on his project management team; they in turn will be reliant on contractors, subcontractors and suppliers who have been chosen for their experience and suitability. However, each member of this greater team is dependent on others and the procedures, systems, contracts and controls must allow the benefits of cooperation to come through.

Such a balance is best achieved through an appreciation of the requirements of other members of the project and it is hoped that this book will help each project member understand how the other disciplines do their work and thereby assist them in achieving the cooperation and efficiency we all strive to obtain.

Owners and contractors involved in industrial engineering projects are constantly seeking new ways of undertaking capital projects in order to increase the efficiency of the industry, to improve the economics of working and to adapt to the changing safety and environment requirements around the world.

Therefore, while every effort has been made in this book to ensure the accuracy of the information presented, the Joint Development Board cannot take responsibility for any personal opinions, errors or superseded information since statutory requirements are constantly changing and practices and procedures vary from company to company.

Any person taking a position of responsibility within an engineering project is strongly advised to acquaint himself with the general requirements of a particular company and the specific requirements of a particular project since no book can cover the multitude of differences in style and definition in use worldwide.

Management of Engineering Projects

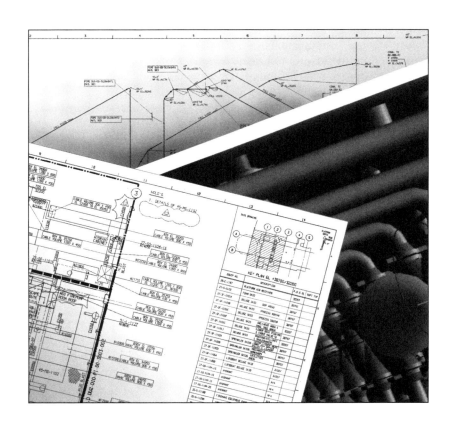

2.1 INTRODUCTION

Project management in one form or another exists in every creative activity involving more than one individual. The term 'management' generally implies organizing people to achieve specified objectives; however, project management has the additional implication of being related to a particular project and hence has a specific start and completion date.

This chapter outlines the role of project management, some of the tools used by the project management team and identifies some of the factors that the project management team have to consider during the course of a major project.

Project management for anything other than the simplest of tasks is not an optional activity; it is a key management activity whose function is to either obtain or use available resources of labour, equipment and material to supervise and achieve defined objectives within stated restrictions of time, cost and quality. This is achieved by formulating and implementing plans to meet the defined objectives, by monitoring performance against those plans and by taking any corrective action deemed necessary.

Management style is peculiar to an individual project manager and influenced by the circumstances in which the manager and his organization are required to operate. The principles of good management practice, however, apply to all circumstances, and while styles may vary the fundamental requirements of good management remain.

In the same way that individuals have a management style, organizations have their own cultures, values and operational policies that govern and influence the way in which they and their personnel operate their business. These cultures, values and policies may:

- be highly commercial;
- have a social or environmental impetus;
- be focused on industrial relations;
- be influenced by external perceptions – the public, financial institutions, etc.;
- be largely driven by statutory or regulatory requirements.

Such matters are subtle, but have a far reaching impact. They must be understood by the management team and addressed in the execution of the project if that execution is to be effective.

Owners will appoint a project manager to take overall responsibility for their project and each major contractor or supplier will appoint a manager, who may also be titled 'project manager', to take responsibility for its share of the project. Although this chapter is generally addressing management by the prospective owner, the majority of the principles described are also applicable to the contractor's side of engineering project management.

2.2 PROJECT MANAGER

The term 'project manager' is in common use in the engineering industries and is the term used in this chapter to describe the most senior position in a project. The term 'engineer' is also in common use, particularly in companies that have an affinity with civil engineering and is a term frequently referred to in many standard forms of contract to describe the person who has full authority to act on all matters under a particular contract.

The engineer when referred to in a contract may also be the project manager, or indeed any capable person nominated to that position. There may be both a project manager and an engineer on a single project; in such cases the project manager may be in overall charge and the engineer be a consultant acting independently on design and possibly contract matters.

It is generally accepted that any project must ultimately be the responsibility of a single person and the activities described in this chapter relate to this 'ultimate' project manager.

Subject to prescribed limits defined by the owner or co-venturers, the project manager will have the authority to decide how the project will be implemented and will typically be given the following responsibilities:

- to discharge the delegated authority received from the owner or co-venturers;
- to act as key focal point concerning the owner, co-venturers and external statutory authorities;
- to report to the owner bodies at specified points;
- to build and lead the project team;
- to set and expedite the project objectives;
- to manage all activities relating to the project;
- to manage any conflict of interest.

Whether the project manager is employed by an owner or by a contractor, he or she must serve the requirements of both the owner and his own project team and ensure that all decisions within his or her authority are made and communicated expeditiously. The project team is entitled to expect the owner and/or project manager to respond with the same degree of urgency to which they are subject.

The project team will provide expertise in the planning, estimating, designing, procuring, constructing, cost control, quality control and commissioning of the plant, but the project manager will look to the owner for a clear directive as to any mandatory requirements that the owner's organization may wish to impose on the project, including:

- provision of the project brief;
- development of the brief, as necessary;
- identification of corporate policies and standards;
- identification of reporting and approval requirements;
- identification of interfaces within the owner organization.

The project manager must be able to motivate the various groups within a work environment and to make personnel aware of and concerned with achieving the overall project objectives and not simply those limited responsibilities that fall within his sphere.

2.3 PROJECT ORGANIZATION

The project manager will build and lead the project management organization that acts as the focal point for the project and manages and oversees the execution of the work. A strong effective project organization is critical to the success of any large venture and the most efficient project organization comprises a team of dedicated professional managers working together from inception to completion of the project.

A strong organization is not necessarily a large organization as strength can come from the close working relationship of a small, highly motivated team. Owners have at times used large project management teams but it is generally considered that such teams can be difficult to control and can result in considerable 'opinion engineering' with resultant cost escalation and lack of time control.

Many owners now wish to see smaller project management teams and while many projects maintain relatively traditional organizational relationships, the latest thinking is to form a single integrated team comprising owner's and contractor's personnel chosen on a 'best for job' basis. This approach aims to avoid the all too common adversarial approach to contracting but clearly affects the contractual relationships between the parties.

2.3.1 Nature of project organization

The following are the five basic types of project organization commonly in use:

- owner based project team;
- owner based project team supplemented by a contractor provided project services team;
- management contractor;
- main contractor;
- alliancing/partnering.

2.3.2 Owner based project team

Where an owner has a unique technology that he wishes to retain in house, or has significant design and construction experience with appropriate personnel available, the owner may opt to provide the whole of the project team (see Figure 2.1). In such circumstances the owner's project team will allocate packages of design or construction work to contractors, will supervise and control the various interfaces between work packages and possibly carry out some preliminary design work in house. The owner's team may also procure items which are critical with regard to delivery or design requirements and provide such items on a 'free issue' basis to a construction/installation contractor. The owner team will be in direct contact with suppliers, designers and construction contractors, etc.

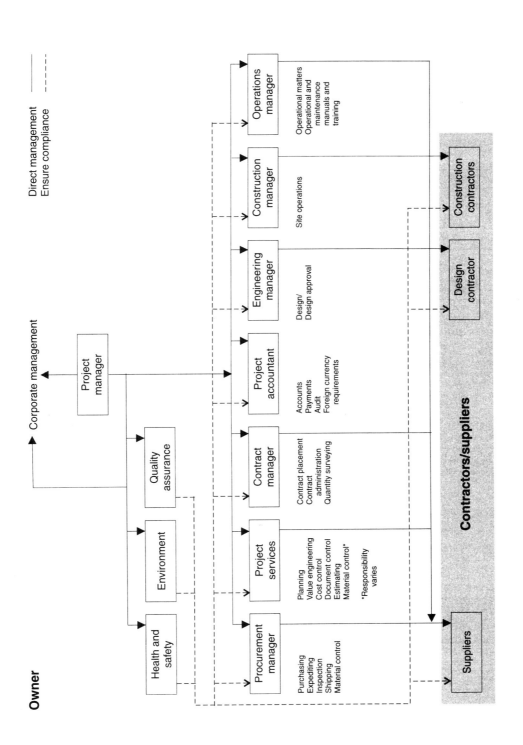

Owner

Corporate management

Direct management ————
Ensure compliance – – – – –

Project manager

Health and safety

Environment

Quality assurance

Procurement manager
Purchasing
Expediting
Inspection
Shipping
Material control

Project services
Planning
Value engineering
Cost control
Document control
Estimating
Material control*

*Responsibility varies

Contract manager
Contract placement
Contract administration
Quantity surveying

Project accountant
Accounts
Payments
Audit
Foreign currency requirements

Engineering manager
Design/
Design approval

Construction manager
Site operations

Operations manager
Operational matters
Operational and maintenance manuals and training

Contractors/suppliers

Suppliers

Design contractor

Construction contractors

Figure 2.1 Typical owner organization

2.3.3 Owner based project team with a project services contractor

Project services comprise those disciplines that are essential to the proper control of time and cost of a project, including document control and material control. This service may be provided by a project services contractor whose scope will typically include those services shown in Figure 2.2. The scope may also, in some cases, include contract management. This type of organization is commonly used where an owner does not have sufficient, or experienced resources to provide a full project team but nevertheless wishes to retain direct control. The intention will be to construct an effective integrated project management team from the complementary strengths of the owner's organization and that of the project services contractor. Packages of design or construction work will be allocated as in section 2.3.2. The project services contractor will be in direct contact with suppliers, designers and construction contractors in connection with those matters for which the project services contractor is responsible.

2.3.4 Management contractor

Under the concept of a management contract the owner contracts the management of the project to a management contractor who controls and coordinates the work carried out by suppliers and design and construction contractors, for and on behalf of the owner. The scope of work may therefore include design approval, planning, cost engineering, procurement, contract engineering, quality control and administration of the project and the various purchase orders and contracts.

The owner/management contractor organization is shown in Figure 2.3. Under this arrangement the owner supervises and controls the management contractor but he has little direct contact with suppliers, designers and construction contractors.

2.3.5 Main contractor

Under the main contractor (see Figure 2.4) approach the owner contracts the entire project, or substantial parts of the project, to a single contracting entity. The main contractor has sole responsibility for the execution of the work, which may include design and which the contractor may undertake using its own resources or those of subcontractors. On a major project there may be a number of main contracts, each dealing with separate and distinct sections of the work (i.e. process unit, tank farm, loading jetty, etc.), in which case the owner may use any of the previous management types to control and coordinate the overall project.

2.3.6 Alliancing and partnering

In traditional contracts the objectives of the contracting parties are diametrically opposed, the owner wishing to hold the contractor to a contract value that might have been based on the minimum of information and the contractor seeking to maximize his profit by aggressively seeking changes

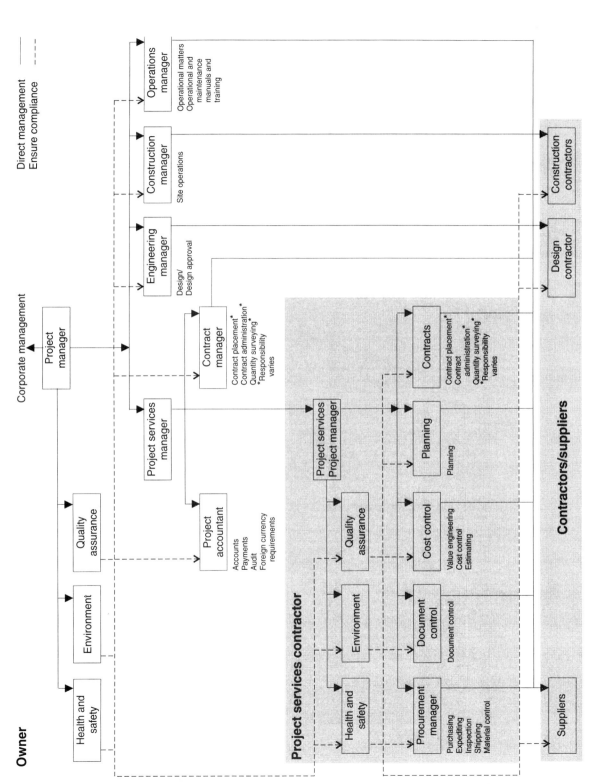

Figure 2.2 Typical owner/project services contractor organization

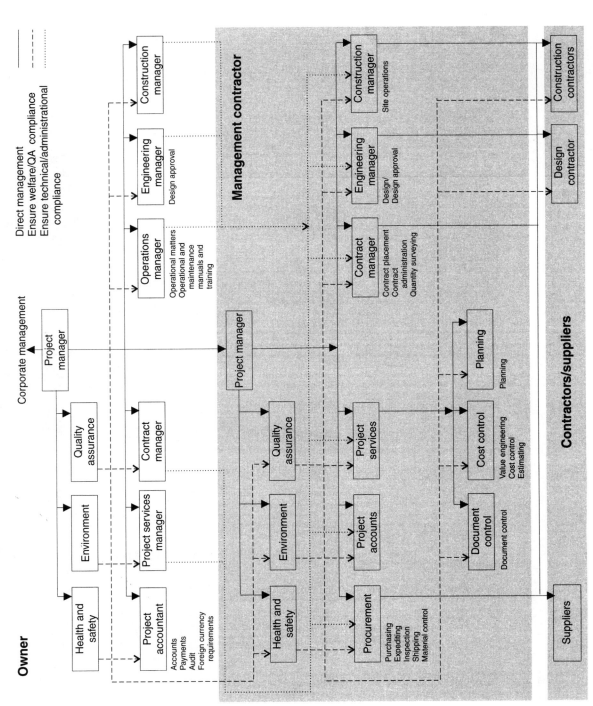

Figure 2.3 Typical owner/management contractor organization

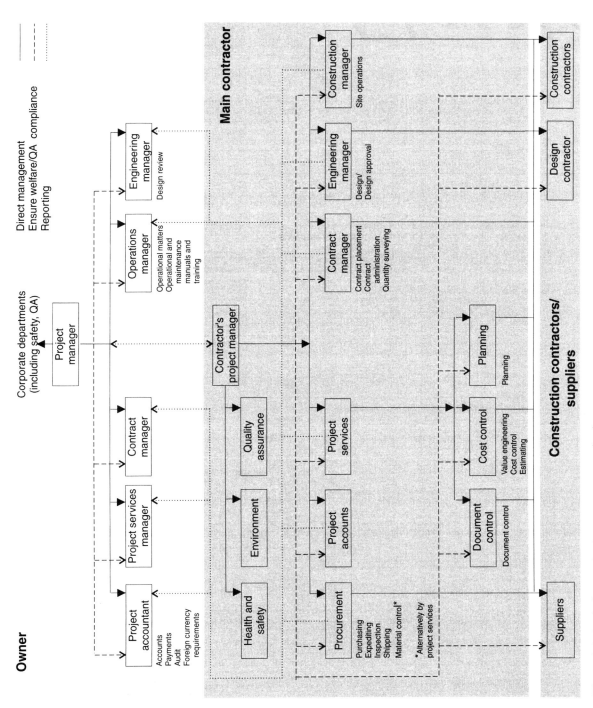

Figure 2.4 Typical owner/main contractor organization

and claims. The net result is a serious diversion of management effort into defending the various contractual positions rather than seeking efficient and economical solutions.

The concept of an alliance or partnering relationship is based on the belief that cost can be saved and profit enhanced when the strengths of the various parties involved in a project are formed into a cohesive team that has the single objective of completing the project on time, within set cost limits and to the required quality.

The resulting organization is a single integrated team comprising personnel from the owner and various contractors chosen on a 'best for job' basis with no distinction made between owner and contractor personnel. It is hoped that such relationships will align objectives and remove confrontation, thereby improving efficiency and reducing costs.

A typical arrangement for such a relationship stresses mutual dependence by reducing contractual risk to a minimum through the use of primarily reimbursable types of contract and linking profit directly to reductions made in out-turn costs. Work subcontracted by the contractor is usually reimbursed at cost, but such subcontracts may be let on a lump sum, measure and value, or on a similar incentive basis to the main contract.

Although the two terms are frequently interchanged, 'partnering' is a term used by some companies to describe an ongoing relationship over a number of projects, while an 'alliance' is formed for a single project.

2.3.7 Selection of project management team

The size, shape and composition of an organization is clearly important, but what is vital is the manner in which an organization allows the various teams within it to work together.

One of the first actions of the owner is to appoint a project manager to take responsibility for the project. The early appointment of the project manager and other key members of the project team should ensure continuity in the important transition from early development to project execution. The project manager will organize and coordinate the various viability studies necessary before the owner makes the final decision to undertake the project. Such studies may, alternatively, be undertaken by the owner's organization or by consultants.

Once a decision to proceed has been taken, the project manager first develops a contracting and management philosophy followed by a preliminary staffing plan and the selection of a project team.

The structure of the project management team depends upon:

- the project contracting strategy;
- the capabilities of people available to form the project team;
- the form of contracts to be used;
- whether responsibility for design is placed with a separate design contractor.

Although the basic shape of the organization will remain throughout the project its content will change from time to time to suit the requirements particular to the stages of the work.

Factors that influence team and individual effectiveness include:

- utilizing the owner's and contractor's strengths;
- minimizing the size of the project team;
- minimizing duplication;
- establishing clear team objectives and individual roles;
- establishing manageable and managed interfaces between account-able persons;
- arranging single point accountability for discrete parts of the work.

One of the key management skills required of a project manager is the ability to identify those members who need to be involved in management discussions and decisions. The familiar statements 'this is not my problem' and 'this problem does not concern my department' must be avoided at all cost. This is particularly so on engineering projects which require very diverse project teams.

There is an inherent possibility for conflict within the project team since each participant sees the project from a particular viewpoint. It is therefore necessary for a project manager to decide upon issues where two or more parties disagree on particular actions and to facilitate a solution to the problem.

2.3.8 Roles and responsibilities

Project organizations need to reflect the matters for which the owner or the contractor is responsible. The various responsibilities of the contractor and the owner should be defined at an early date and the nature of the various interfaces identified throughout the organization.

Considerable time can be wasted and expense incurred through poor coordination. Where there is no single management responsibility for an interface between, for example, design disciplines or between procurement and construction activities, a coordinator should be identified and given the responsibility for ensuring that communication between the parties at the interface is effective.

A schedule of all the interfaces and the responsible coordinators should be prepared during the early stages of the work and reviewed as the project develops.

2.3.9 Personal objectives

The project execution plan described in section 2.6 describes the owner's need to focus the project on specific objectives. Groups within the project organization and individuals within the groups need to be able to identify their own objectives and to understand how their personal objectives link through the group objectives into those adopted by the project.

All members of the project management team should be identified against positions shown on a project organization chart with a clear description of their roles and responsibilities. Such roles and responsibilities should not duplicate those of other members of the team.

Most problems start at the detail level and most solutions can be found there. A team that has clear objectives, communicates freely and gets the members fully involved will build up a degree of trust that allows for early identification of problems and their correction.

Project team members should be empowered with the responsibility and authority for achieving the project objectives in their respective areas of the work. Empowerment means accountability for ensuring that a given task is completed as required. Empowerment does not mean that the accountable person is himself responsible for undertaking the various specialist tasks necessary to carry out the activity.

2.3.10 Confidentiality

Industrial engineering projects invariably involve the manufacture of a product to meet a market need and make a profit. The economics and means of manufacture are therefore commercially sensitive and it is frequently the case that contractors and subcontractors are required to sign confidentiality agreements covering a broad range of information and documentation produced in connection with the project.

Such agreements may cover not only information provided by the owner but also information produced or developed by the contractor, subcontractors and suppliers, on behalf of the owner.

2.3.11 Operations support

A representative of the owner's operating company or division should be assigned to the project team as operations coordinator to advise on operational issues through early input to the design and to maintain continuity of support through commissioning and into the operation or manufacturing stage.

It is essential that a designer fully appreciates the end user's requirements of the facility before undertaking design work. Operations responsibility therefore spans not only that of the traditional 'owner' in ensuring that the finished facilities perform the intended function, but also involves the input of experience of operational issues to ensure that the design takes due account of the need for safe, efficient operation and maintenance. These issues include engineering, operational maintenance, warehousing, product and waste management, information technology, logistics, etc. Organizing this variety of inputs within the framework of a project needs careful consideration, particularly with regard to timing.

Early input to design at a time when the cost of change is at its lowest is to be preferred. Operational input should therefore be made available before the commencement of detail design. Many examples exist of operations personnel requiring extensive changes to be made after seeing the proposed facility for the first time at handover and of operations groups having reviewed designs immediately before issue for construction and requiring change, with consequent delay or disruption to construction activities. Late involvement reduces the benefits that the operations group can bring to a project, significantly increases the cost of any required change and is the principal cause of conflict between project and operations groups.

Operations groups typically have the following responsibilities:

- to provide single point contact for the project team on all operational issues including the effective operation of emergency services, etc. and to ensure the effective transfer of such information to the project team;
- to consider all issues relating to the operation of the facility including operability, maintainability and logistics;
- to ensure timely input of operational experience and functional requirements into the design, in order to provide facilities which minimize safety hazards and operating breakdown and maintenance costs;
- to provide guidance and resources on the testing and commissioning of the plant;
- to assist the project team in the performance of hazardous operations analysis, quality audits and safety case reviews;
- to provide information on the limitations and constraints from adjacent or interconnected plant operations;
- to provide information on existing site conditions including underground services, disused foundations, etc.;
- to provide information on sensitive local issues of which the designer or contractor may be unaware;
- to identify requirements for the storage or warehousing of incoming supplies, outgoing product or by-product, waste, etc.;
- to develop and coordinate all necessary operation and maintenance training and documentation;
- to review and approve operations and maintenance manuals;
- to consider any issues related to eventual decommissioning.

2.4 OUTSIDE INFLUENCES

The project must gain an understanding of the influence and potential effect of governmental and market conditions and of the environmental lobby; these can all have a substantial impact on the success of a project. Additional factors may be the impact on the project of other projects competing for local resources, or product market competition, where being first with a new product, or increasing production of an existing product, is all important.

The project team must obtain this understanding through discussions with the groups involved and by gathering and analysing data on resources and completed projects that provide an insight into future trends.

2.4.1 Government influence

A major effort must be directed towards obtaining and maintaining accurate information on legislation, regulations and documentation requirements. It should be remembered that fiscal policy changes, budget changes and changes of government may affect the commercial viability of a project. Many of these aspects will be considered before the owner makes the decision to commence with the project.

The Department of Trade and Industry (DTI), the Health and Safety Executive (HSE), the European Union (EU) and local government councils as appropriate, should all be contacted and the contact maintained to identify requirements and any changes in requirements, that may affect the project.

In particular the project team must be fully aware of all matters requiring consent or approval by outside bodies, since approval requirements can have a major impact on:

- programme;
- engineering resources;
- outside specialist support;
- scope changes and cost.

2.4.2 Resource markets

By comparison with the building and civil engineering sectors the number of owners and contractors in the engineering industry is quite small and there are certain key areas in which a near monopoly exists.

To avoid such problems the project must identify and keep track of any critical resources and determine the number and type of projects that are competing for scarce resources in the same geographical area.

Many major engineering projects are sited in remote areas. Local resources may not be available in the required numbers, or may not possess the requisite skills, thus involving the importation of labour or heavy plant, the construction of accommodation for the imported labour or perhaps off-site prefabrication of parts of the plant.

A review of labour resources, customs and practices and labour and management relations in those areas of anticipated project activity should be maintained.

2.4.3 Patents and licences

The method of production, be it a chemical process or manufacturing technique, may in whole or part be the subject of a patent taken out by its inventor. Projects which involve such patented processes will require the owner to seek a licence for their use, and in addition the design contractor may be required to subcontract that part of the design which relates to the patented process to a designer who either owns the process or who is itself licensed to use the design.

It is often the case that owners will make it their contractual right to claim ownership of 'inventions' made during the design and development of their project.

2.5 CORPORATE AND CO-VENTURER'S REQUIREMENTS

Although a project team may be seen as a 'stand-alone' organization, it will still be required to conform with certain requirements of the owner or co-venture partners.

Any venture that is undertaken in association with partners will call for special arrangements in communication, reporting, decision making, approval and funding. The obtaining of approvals is the key to maintaining progress on a multi-owner project and the project team should establish, at an early date, the information required by corporate departments of the owner's organization and co-venturer partners for project approvals and their anticipated timing. These groups should be contacted early and often, since their needs are constantly changing and any changes must be monitored. Where possible, early review and approval of the project's strategy and concepts should be obtained which will help prevent redesign and delays.

Items to be approved may include:

- design philosophies;
- procurement and contract strategies;
- approvals for expenditure;
- design solutions;
- development plan and budget;
- contract plan and bid lists;
- contract awards.

Co-venture partners should be contacted early in the project to address:

- what they wish to approve on the project;
- what information is required for approval;
- how much time is required for approval;
- what are the reporting requirements.

Not only will co-venturers require to be convinced of the viability of the project before authorizing expenditure, but they may also require to approve funds for each stage and to be kept informed of all major project activities and given reasonable access to the project team.

Co-venture partners and corporate departments outside the project should be requested to give a very high level of contracting authority to the project team. The project team may, for example, develop a contracting plan during preliminary engineering that specifies bid lists, contract type, approximate value, timing, evaluation requirements, etc. Approval by the corporate departments and co-venture partners, with the understanding that if the contracting plan is followed the project team will not be required to request intermediate approvals, will allow the team to keep the project moving at a steady pace without the wait for further approvals. Such authorization will not preclude auditing or otherwise checking to ensure that the project is progressing in accordance with the approved plan.

2.6 PROJECT EXECUTION PLAN (QUALITY PLAN)

Due to the sheer size of many engineering projects, the number of persons involved and the variety of their contributions, the necessity for teamwork has to be emphasized and an environment created in which an efficient team can evolve.

The project execution plan is the pinnacle of the project quality system; it states the broad requirements of the project and translates these into a detailed and coherent strategy. In addition, it initiates other documents that expand, implement or show the results of what is described in the execution plan.

Project execution plans are produced by the project team to cover its main activities. Similar plans will be produced by each major contractor for their activities that will relate to the project execution plan prepared by the project team.

Quality assurance is covered in more detail in Chapter 6 on quality assurance.

2.6.1 Development of the project execution plan

The draft project execution plan is normally developed during the conceptual stage with the final version being issued before commencement of detail design. This states the main objectives of the project team and the strategies, project management processes and activities necessary to the achievement of these main objectives. Generally it will not state how the processes or activities should be carried out, as these will be developed subsequently in separate more detailed documents.

The project execution plan should reference detailed planning documents such as the procurement and contract plan, organization charts, etc. These describe how the project is to be developed, the resources required, the timing and so on, and should be structured so that the project can measure its performance against the project execution plan.

The project execution plan is the main strategy document for the project and should outline the basic structure of the project including:

- project objectives;
- project management;
- project organization;
- project programme;
- project control strategy;
- quality plans;
- safety and environmental policy;
- procurement strategy;
- contracting strategy;
- budgets;
- resource strategy.

Where a company has adopted mission and vision statements, the project execution plan will also contain such statements that are relevant to the project.

Organization charts within the project execution plan should identify the positions to be filled, their roles and reporting links. Individual and group responsibilities and authority levels should be clearly defined as should the divisions of work and interfaces. Detailed job descriptions and individual scopes of work may also be required.

2.6.2 Project objectives

One of the major factors in establishing a good team is to focus attention on the desired result by providing an understanding of what will be considered to be a successful project.

The project objectives are a statement of what must be achieved if the project is to be considered a success. The project objectives should provide a focus towards the fundamentally important aspects of the project and should either be a measure of its success or should encourage behaviour leading to that success. Typical project objectives relate to operational requirements, project completion milestones, safety goals, product quality and cost. Project objectives should be reasonably small in number, such that each one can be remembered and be seen as important. They should consist of clear, unambiguous, memorable simple statements that can be identified by the project team and those who support it. Objectives should be realistic and achievable; the inevitable failure to achieve impossible objectives rapidly destroys morale and quickly disillusions the team.

2.6.3 Mission and vision statements

Some owners and contractors adopt mission and vision statements for their company or project above the level of project objectives. Such statements comprise a short, sharp, precise statement of intent that acts as a focus towards which the objectives and critical success factors are directed.

The mission statement is the 'bottom line' and is a statement of what the company, or project is seeking to achieve in 'hard' terms, i.e. 'to do something by a set date or within a stated cost'. The vision statement is a 'vivid picture' of an ambitious, desirable future state that is better in some important way than the current state, i.e. 'to be seen by our customers as being best in class'.

2.6.4 Critical success factors

Critical success factors are those which describe how the project objectives will be achieved. Some critical success factors which are common to all projects include, for example:

- obtaining input from the operations group prior to commencing detail design;
- provision of a safe working environment;
- development of a high quality integrated project management team;
- completion of design of a section before commencing construction.

Each project will have unique circumstances dependent upon the primary objectives of the project which may make some commonly accepted factors inappropriate.

2.6.5 Project programme and milestones

Planning is the sequencing, scheduling, monitoring and reporting of events, activities, milestones and resources and provides the datum against which the project status, progress and changes can be measured.

The project programme incorporated into the project execution plan will ideally be the level 2 programme. Milestones or target dates identified in the project schedule mark critical events such as key decisions or approvals, or important interfaces between dependent activities and focus special attention on the project pressure points.

Planning is covered in more detail in Chapter 5, Part 2 on planning.

2.6.6 Procurement strategies

Since a high proportion of the project funds are expended on materials and contracted work, the development of a sound procurement and contracting strategy is one of the most critical elements in the project execution plan.

Procurement is usually controlled against a procurement plan that identifies all materials and equipment to be purchased by the party producing the plan and possible sources of supply. Particular reference will be made to items which come under the requirements of EU and UK regulations governing competition and tendering. The procurement plan is also a document against which procurement activities are monitored and controlled.

2.6.7 Free issue equipment and materials

Circumstances on a project may dictate that certain materials (a term intended here to include equipment) are to be issued to the contractor by the owner on a free issue basis, notably vessels, electrical and mechanical equipment, instruments, etc. together with bulk items such as pipework, electrical cable and the like.

The decision to provide materials on a 'free issue' basis is made for various reasons including:

- design information being required from the specialist supplier in order to complete the overall design;
- items being of a specialist nature and requiring very exacting standards of performance and specification;
- items being on an extended delivery and the contractor being unable to obtain the materials in a period anticipated by his construction programme for that element of the work;
- the buying power of the owner;

- the requirement to match materials or equipment already in use on an owner's existing installations;
- the owner already owns or produces the material.

Where a supplier is required to design the equipment, agreement must be reached as to the documentation to be produced, the programme for its production and the approval cycles required. Such information is vital to the principal design effort and interfaces between the supplier and designer must be managed effectively if the design is to proceed in an efficient manner.

2.6.8 Supplier (vendor) engineers

Specialist equipment may require specialist assistance during installation and commissioning in order to ensure its correct operation and to ensure that any guarantee regarding performance or quality is preserved. This requirement for supplier (or vendor) engineers must be identified early in order that it can be anticipated in purchase orders for equipment supply and in contracts which include the installation and commissioning of the equipment. The appreciable costs involved must also be included in budgets and the procedures produced for their control.

2.6.9 Contracting strategies

In the context of this chapter a contract strategy is the breaking down of the project into packages of work and deciding on the most suitable method of payment for each package.

The nature of the work packages is dependent on a number of factors including:

- length of the overall construction programme. Is it necessary to commence site clearance ahead of completion of the main design?
- similarity of the work with other parts of the project. Are there a number of similar 'structures', how do they relate geographically and in terms of programme, can they be combined into a single package?
- single point responsibility. Is it possible to avoid dispute by making a single contractor ultimately responsible for the whole?
- status and availability of design at enquiry and award stages and during the construction period. Is a lump sum contract possible, is availability of design likely to cause ongoing problems of delay and disruption?
- early specialist input to design. Is it necessary to appoint a contractor ahead of design being available in order to provide an input to the design or advise on constructability, etc.?
- benefits of scale. Will the 'overhead' costs of site management and plant utilization, etc. for a large package produce more economical results than a number of smaller packages?

Once the nature of the various contract work packages has been decided the type of contract will be reviewed.

The type of contract and the manner in which the contractor is to be paid depends on the nature of the work and on the status and availability of design and materials at enquiry, at award and during the construction period – can a lump sum be requested or will the likelihood of interference (change, delay disruption, etc.) from other parallel contracts, the owner or his designer, require the use of a reimbursable form of contract?

The control of the preparation, bidding and award of contracts is against a contracting plan which is similar to the procurement plan in that it identifies the work to be contracted (or subcontracted) by the party, together with information relating to programme requirements and to the anticipated value and type of contract to be used, e.g. lump sum, measure and value, etc.

As with the procurement plan particular reference is made to items which come under the requirements of EU and UK regulations governing competition and tendering.

Prospective contractors may be listed in the contracting plan and any screening requirements identified, with an allowance made in the programme for this pre-qualification process.

2.7 PROJECT STAGES

Historically, projects have been broken down into a series of stages. Owners use various terms for the stages of a project and each project develops its own definitions of the required stages, which may typically include:

1. design:
    - feasibility stage;
    - conceptual design;
    - preliminary design;
    - detail design and procurement;
    - design deliverables;
2. construction;
3. initial operations:
    - pre-commissioning;
    - commissioning;
    - initial start-up;
4. operation.

A project may choose to combine or subdivide the above stages to suit its particular requirements.

2.7.1 Feasibility stage

The feasibility stage of a project will comprise studies undertaken on a very generalized basis to evaluate the technical and commercial benefits that the plant will provide to the owner.

2.7.2 Conceptual design

Conceptual design may be carried out in house by staff engineers perhaps supplemented by temporary personnel or by specialist design contractors.

Although the greatest impact on final cost can be made during the conceptual design phase, owners and their advisers too often plan this phase poorly and allocate insufficient time and financial and human resources to these vital early stages of a project. Pressure to initiate a new project quickly can result in subsequent inefficiencies and costly errors.

During the conceptual design stage any new technology is identified, studies carried out, outstanding information incorporated and the design philosophy developed. The decision to incorporate and use new technology can greatly affect a project and must be a balance between potential cost savings attributable to new technology and the risk of likely cost overruns due to their development.

The purpose of the conceptual design stage is to:

*   confirm that all the engineering solutions proposed are technically and economically feasible;
*   ensure all critical technical issues have been identified and addressed;
*   seek limited planning consents;
*   plan preliminary engineering.

2.7.3 Preliminary design

The details of the conceptual design are reviewed by the team and their preferences defined and incorporated, or otherwise eliminated from consideration.

The operations group's major contribution to the project will occur during preliminary design. During this phase the operations group should review and approve all deliverables. As the project progresses through preliminary engineering the results of the feasibility and conceptual design stages are incorporated into equipment and package specifications and the basis of design produced for detail engineering.

Constructability of the facilities is verified during preliminary design. Contractors and especially offshore module fabricators have different methods or sequences of construction. Where constructability is likely to be an issue, potential contractors should be invited to review the design and their comments considered for incorporation therein.

In certain extreme cases, i.e. where a specialist technology is required or where early completion is paramount, it may be necessary to select a contractor before the completion of preliminary design. A contract awarded early in the design stages will not have a well-defined scope of work and is contrary to the traditional and recommended practice of awarding contracts based on a fully developed work scope. The solution to this dilemma may require some innovative thinking in the contracting strategy.

The purpose of preliminary design is to:

- mobilize the project team, design team and project services;
- conduct studies to optimize and define design requirements;
- increase the accuracy of planning and cost information;
- conduct safety and environmental studies;
- develop process and instrumentation diagrams, single line diagrams and layouts;
- investigate constructability;
- confirm new or innovative technology;
- arrange provision of supplier data, and place purchase orders as necessary to meet the programme;
- seek approvals and planning consents.

Management of additions and changes becomes more important as the design progresses and the project team will use a formal procedure of change control during preliminary design. The increasing importance placed on the control of changes to the design, their cost and effect on the programme, serves to emphasize why options should be considered in full during the earlier phases.

2.7.4 Detail design

Change control becomes even more critical during detail design. Changes should be avoided wherever possible, but if changes have to be made, they should be challenged and if considered acceptable should be incorporated as soon as they are discovered so that their effect can be minimized. Changes may include not only changes to the design, but changes to programme, cost, delivery times, etc.

The activities during detail design are to:

- develop details and finalize drawings and specifications for construction and commissioning work;
- procure materials, plant and equipment and construction contracts;
- ensure design complies with any safety case requirements and environmental assessment;
- obtain certification agreement.

Detail design will frequently be undertaken by an engineering contractor, with the owner's role being to stipulate project requirements, to approve specified key aspects of the design, to identify the operational requirements and to manage the interfaces with outside groups such as co-venturers and government agencies. The owner's role should not be to mark the contractor man for man, or to check the contractor's work in detail, or to lead the design process. The owner's role should rather be to ensure that his requirements relating to quality, performance, operability and safety are met, while allowing the engineering contractor to use his experience and expertise in the efficient design and construction of a complex project. This requirement makes it essential that experienced and competent contractors are selected.

2.7.5 Design deliverables

Each stage of a project and each work package within a project will be required to produce something tangible, i.e. layouts, general arrangements, specifications, enquiry documents, equipment and materials on site, process guarantee test certificates, etc. The physical results of design and other management activities are frequently referred to as 'deliverables'. It is the deliverables that provide the basis for reporting project design status and for recognizing achievement. The main documents or deliverables resulting from the various design stages are covered in Chapter 3 on estimating.

2.8 CONSTRUCTION

Construction will be carried out by one or more main contractors each of which will employ various specialist subcontractors. Responsibility for coordination between the main construction contractors will rest on the project organization, as chosen by the owner and headed by the project manager.

Expenditure increases rapidly during the construction stage due to the high levels of labour, plant and materials involved in the work. Any disruption to the efficient use of these resources will result in significant cost overrun. Serious damage can be done to the project programme and budget by change, delay and disruption. However, the project manager must be vigilant that attempts to increase progress or reduce expenditure do not jeopardize safe working practices or the quality of work performed.

The project manager will rely on the project team to keep him appraised of progress and cost and will scrutinize weekly or monthly reports, hold regular progress meetings in order to resolve current difficulties and anticipate and resolve anticipated logistical and contractual problems. The project manager will also wish to ensure that sufficient records are being maintained by the team to safeguard the owner's contractual position and to satisfy the requirements of the statutory and certifying authorities and the eventual operators of the plant.

The project manager should insist that progress reports do not simply report progress achieved, but that they deal adequately with activities that have not happened, since it is these 'negative' aspects of the report that provide early warning of future problems.

Completion of construction work typically means that the contractor has erected the plant in accordance with drawings and specifications, completed any specified pre-commissioning work and completed the final clean up. However, as-built drawings, operations manuals and handover of other specified documents will be required before the full contract is deemed complete.

2.9 INITIAL OPERATIONS

Initial operations is a term used to describe the entire process of pre-commissioning, commissioning, initial start-up and production and performance testing of a plant.

Various terms are used by different companies to cover the activities required to fine tune a plant and bring it into production and the terms and definitions used here are indicative only. The sequence of activities leading to commissioning and acceptance of a process plant will typically be as follows:

Construction

⇓

Pre-commissioning

⇓

Mechanical completion

⇓

Commissioning

⇓

Mock operation

⇓

Initial start-up

⇓

Check product specification

⇓

Check production performance

⇓

Acceptance of plant

2.9.1 Pre-commissioning and mechanical completion

Pre-commissioning activities are those which have to be undertaken prior to operating equipment such as adjustments, cold alignment checks on machinery, etc. performed by the construction contractor prior to commissioning and without which the installation cannot be said to be mechanically complete.

Mechanical completion of a plant or any part of a plant occurs when the plant or a part of the plant has been completed in accordance with the drawings and specifications and the pre-commissioning activities have been completed, to the extent necessary to permit the owner to accept the plant and begin commissioning activities. Mechanical completion will typically be deemed to have been achieved on completion of the following:

- hydrostatic testing of all pipework and equipment;
- testing of electrical systems and specialist equipment packages;
- cold alignment of machinery;
- loop checking of instruments;
- flushing of pipework;
- acceptance of all documentation in accordance with the procedures, drawings and specifications.

2.9.2 Commissioning

A plant, or any defined part of a plant, is ready for commissioning when the plant has achieved mechanical completion.

Commissioning activities are those associated with preparing or operating the plant or any part of the plant prior to the initial start-up and are frequently undertaken by the owner or a joint owner/contractor team. Commissioning may involve mock operations which are commissioning activities conducted to allow operational testing of the equipment and operator training and familiarization.

At the completion of commissioning, the plant will be fully ready for production operation.

2.9.3 Initial start-up

Initial start-up occurs when feedstocks are introduced to the plant for the express purpose of producing a saleable product for the first time, often referred to as 'oil-in' in the hydrocarbon processing industry.

2.10 PROJECT RISK ANALYSIS AND MANAGEMENT

2.10.1 Risk analysis

Extreme happenings, such as *force majeure*, cannot be ignored when making a decision to sanction a project. Although on a straightforward project such risks can be considered without elaborate calculation, on a major project the degree of uncertainty connected with cost, time, location, technology and quality and the complex manner in which the various risks interact to magnify or extinguish each other, can only be fully explored, substantiated and presented by undertaking a comprehensive analysis of the full range of risks involved. The very process of undertaking the analysis provides significant help in minimizing risk since it focuses the project team on those areas requiring greatest attention.

Risk analysis is a tool that provides a formalized approach to the identification and evaluation of a complex series of uncertainties, their probabilities and interdependencies and provides a logic and focus for the continuing management of risk. Because risk analysis includes in its assessment the probability of the event occurring and presents this information in a manner from which the decision makers can undertake their cost and benefit analysis, it is capable of handling provisions against all types of risk, including *force majeure* and other extreme eventualities that would not normally be included in a traditional contingency allowance.

The increasing use of computers has encouraged the development of sophisticated techniques for the analysis and management of risk on construction projects. While these techniques are of undoubted value to a project, their use has to be anticipated at inception since it requires amendments to the way in which decisions are made and the time required to undertake the analysis must be allowed in the programme established for the project.

As with a traditionally calculated contingency allowance, the provision for risk can be a very significant amount within an estimate and not only deserves to be handled in a manner appropriate to a major cost element, but also requires the analyst to be able to explain the basis on which risk was calculated and the probability of likely occurrence of the subject event.

2.10.2 Identification of risk

The first step in solving a problem is knowing that the problem exists and in risk management the most important phase is the identification of areas of potential risk. The various risks can initially be identified by:

- reference to a checklist;
- subjecting the project team to a brainstorming session;
- collating lists prepared by individuals or departments;
- the analyst from personal experience.

2.10.3 Evaluation techniques

There are three types of risk analysis in common use:

1. elementary;
2. sensitivity; and
3. probability analysis.

Each type of analysis offers differing degrees of sophistication and each corresponds to the particular requirements of the project at the time of analysis.

Elementary risk analysis

Elementary risk analysis is based on the judgement and experience of those undertaking the analysis and simply comprises the subjective judgement of the effect of the most obvious major risks anticipated on a project. Elementary risk analysis is normally used on an uncomplicated project or at the appraisal stage of a major project.

Sensitivity analysis

Sensitivity analysis considers the impact that varying degrees of a single element of risk may have on the project, by analysing and evaluating its effect on the programme or budget assumptions.

A sensitivity analysis is frequently presented graphically, or as a series of figures, but its weakness is that it treats each element of risk individually and does not pay attention to the complex impact on parallel, or later, activities. Sensitivity analysis is frequently used later in the project life when evaluating tenders and considering the impact on cost of design or time changes, etc.

Probability analysis

A probability analysis will take account of the many elements of risk within a project and will apportion a degree of probability and range of values to each. In this way an assessment can be made of the most probable result, with statistical spread of other results. In its most sophisticated form a probability analysis addresses the consequential impact on the ongoing activities and risks.

Although a summation of the most probable results can be made, it is unlikely that in practice the most probable result will be achieved against each element of risk; who is to say that they are the most probable results? To an extent it is a lottery which risk occurs and to what degree. The power of the computer may therefore be used to generate a large number of random scenarios, based on the assessed probability factors, to provide a most likely final result. It is not surprising that this technique is referred to as a Monte Carlo simulation.

2.10.4 Assessing probabilities and values

In order to provide the basic information for the analysis, design engineers, cost engineers, construction managers and others experienced in their disciplines may be asked to accept or amend a list of possible risks, to state a base assumption and the likelihood in percentage terms of achieving that base. The risk analyst will then discuss with the engineer the most applicable shape of distribution covering the spread of possible results before feeding the information into the computer. The risk analyst will use the planning network to consider the impact that the various categories of risk will each have on the other, while taking care not to duplicate risk by including an individual risk that is partially covered by other risks. An example is the risk of time overrun in the fabrication of a part of the work, which may be identified as such in the analysis; but part of this risk may be due to delay in delivery of materials, which may be a separately identified risk item.

A probabilistic risk analysis will produce values and related confidence levels on an 'S' curve, ranging from the certainty that the project cannot be completed below a stated figure and total confidence that it will be completed within a further figure. In the mid-range of the 'S' curve there will be a range of projected final values, each with a related confidence level. The decision to proceed with a project, or its next stage, may be taken at a budget figure that has a confidence level of 50%, i.e. where there is an equal chance of underrunning or overrunning the figure. Alternatively, and depending on the expected return or other benefit that the project will provide, the decision may be taken to sanction the project at a lower budget, but with a low level of confidence that this low figure can be achieved, or vice versa.

A full probabilistic risk analysis can therefore provide decision makers with a greater understanding of the possible result of the decisions taken and means of judging more accurately the greater risk that may be acceptable on a lucrative project.

2.10.5 Risk management

One of the main responsibilities of the project team is the management of risk. Risk can be managed by transference of risk to others, by avoidance of risk, or by controlling the factors causing the risk.

Transfer of risk

Risk can be transferred by placing the responsibility for its outcome on others, e.g. by utilizing insurance or by requesting fixed prices rather than paying for increased costs as they occur. Risk is therefore usually transferred at a cost.

Avoidance of risk

Where risk is due to a factor that can be changed the opportunity exists to avoid the risk by changing the factor, e.g. by changing from new untried technology to a tried and trusted alternative, or moving a sensitive operation from winter to summer.

Risk is therefore usually avoided by accepting what may be seen as a less attractive project option.

Control of risk

The first step towards the control of risk is the identification of the major risk areas and their initial evaluation followed by continuing control. This control is exercised through the normal project control procedures of regular progress measurement and comparison against a defined base, the identification and implementation of required corrective actions and the monitoring and reporting of results of the action taken.

Risk register

Once identified the various risks will be incorporated into a risk register which typically contains the following information against each risk item:

- the likely impact of the risk in terms of cost;
- a brief statement regarding the anticipated means of mitigating the risk;
- the name of the person or department responsible for mitigating the risk.

As the project develops, new risks will be identified and earlier risks will change in status. The project team will therefore be invited periodically to re-evaluate old risks and to identify new ones for incorporation into the register.

Once the risk register is issued the responsible person will use it to prioritize his attention on those risk items which need to be placed into 'intensive care' in order to keep their effects to a minimum.

2.11 SAFETY, ENVIRONMENT AND QUALITY ASSURANCE

Safety, environmental and quality requirements need to be established and planned very early in the life of a project.

Safety is important both in construction and operation; it is the subject of extensive legislation which is covered in detail in Chapter 12 on health and safety. Safety has to be planned, in terms of safe design and in the choice of contractors who are capable of working safely.

Environmental protection policies are under constant review, which makes prediction of requirements in this area that much more difficult and far more project specific.

Considerable attention is therefore necessary not only to get the safety and environmental matters right, but to comply with the changing legislative regime.

It is essential that a manager who is knowledgeable of the legislation is made responsible for these concerns on a project. Such a role may well conveniently include the quality assurance role as they are closely inter-related.

2.11.1 The Construction (Design and Management) Regulations 1994

The accident rate in the UK construction industries has historically been high when compared to other industrial sectors and the Construction (Design and Management) Regulations (CDM), which apply to most design work and construction projects, were introduced by the Health and Safety Executive to improve on the performance of the industry. They came into force on the 31 March 1995.

The regulations proceed on the basis that health and safety responsibilities in construction should be coordinated and shared among all the parties involved, be they owner, designer, contractor or subcontractor. These regulations, which are described in detail in Chapter 12, are frequently referred to by the acronym CDM. They apply to the design and construction of projects including demolition and affect the various persons and organizations associated with them including owners, designers, contractors, subcontractors and suppliers.

The regulations concern the management of health and safety; while they do not apply to every project or everyone all of the time, most projects and people involved in projects will be affected.

2.11.2 CIMAH and Safety Case Regulations

The *Piper Alpha* and Flixborough disasters resulted in regulations being introduced that have prompted considerable change to the design and operational requirements of projects. Regulatory requirements such as CIMAH (Control of Industrial Major Accident Hazards) for onshore projects and SI 2885 (Safety Case Regulations) for offshore projects continue to be developed and the EU is actively producing safety directives that will eventually become UK legislation.

The CIMAH and Safety Case Regulations are similar regulations in that they require the preparation and delivery of a concept and pre-operations safety case for a project. Fundamental to this safety case will be the ability to describe the management system that is in place and which ensures that 'the installation will be designed, selected, constructed and commissioned in a way that will reduce the risks to health and safety to the lowest level that is reasonably practicable'.

The regulations are legal requirements and should contain the information necessary for the Health and Safety Executive (HSE) to undertake a safety audit.

The objectives of the safety case are:

• to describe at the highest level, how the project management team manages safety;
• to describe hazard identification and hazard mitigation and to provide the quantitative demonstration of endurance standards and the total installation risk;
• to describe how a facility is or will be built and to describe the plant and equipment, emergency systems and the process arrangements;
• to describe, in the project development phase, how project safety is managed;
• to describe, in the operational phase, how the installation safety management system operates;
• to verify that the safety case meets the requirements of the proposed safety case regulations.

Design will be demonstrated as being safe by use of qualitative and quantitative tools, sound engineering practice, experience and verification including quality assurance audit.

Project safety must continue with a strategy that ensures that safety is incorporated into every stage of project work, from concept to decommissioning. By this strategy, all options for development will have been designed to be safe and demonstrated as being safe. No option will be recommended for further development that prejudices the safe operation of the plant.

2.11.3 Safety plan

The strategy necessary to meet the requirements of the safety regulations will be embodied in a safety plan that drives the development of the major safety activities. The safety plan has the purpose of detailing the major safety activities (studies and responsibilities) at each phase of project development. As each new stage is reached the safety issues become more specific, the safety plan will, therefore, develop with the project.

The objectives of the safety plan are:

- to indicate the major safety activities;
- to define group responsibilities;
- to initiate studies and monitor the work;
- to assess studies and disseminate the conclusions, recommendations and actions.

2.11.4 Safety philosophy

The safety philosophy is a high-level document and is the starting point in describing how safe design, construction, operations and maintenance will be achieved.

The objectives of the safety philosophy are:

- to provide clear statements and direction for the development of 'safety requirements';
- to ensure that safety is incorporated into all design work;
- to initiate positive thinking of hazard elimination;
- in the case of offshore installations to initiate the provisions for evacuation escape and rescue (EER), and temporary safe refuge (TSR).

The philosophy should be hierarchical in approach and should examine safety to people from the global issues through position, orientation, layout, equipment and safety systems.

2.11.5 Safety requirements

The safety requirements are established by adding detail to the safety philosophy and will be a document that will be used by the design contractor and the project team to monitor safety design work. The objectives of the design safety requirements are:

- to provide a document which in clear tangible terms details the safety requirements for the planned installation;
- to provide a springboard from which individual discipline engineers can take responsibility for the safety requirements specific to them and to develop them into the design.

2.11.6 Environmental impact assessment

The environmental impact assessment, where required, provides management and statutory bodies with a formalized and standardized means of ensuring that the installation performs within legislative and corporate environmental standards.

It assesses the total environmental impact of potential discharges, including the degree of risk and potential severity of accidental release of pollutants. It also addresses the relevance, practicability and effectiveness of plant and equipment, procedures and personnel in the prevention of environmental damage and the mitigation of the effects of all discharges whether accidental or operational.

Reference should also be made to the approvals required under the various statutes as stated in Appendix C.

2.11.7 Quality assurance

Quality management applies to all products and all services and an accredited quality assurance/quality control system is becoming increasingly necessary in order to demonstrate that a company is capable of achieving a consistently high quality of work.

The objective of the quality control process is to provide a system that supports the quality assurance plan for the project, and ensures compliance with regulations covering the installation. These activities are described in detail in Chapter 6.

2.11.8 Certification requirements

The certification requirements for onshore and offshore projects vary considerably. The project should develop a project specific certification and regulatory requirements schedule listing all the requirements, together with details for the appropriate legislative body or representative and lead times for approval submissions.

2.12 LATHAM AND CRINE

Questions have been raised in a number of quarters concerning the efficiency of the UK construction industries, which have resulted in the Latham Report, which primarily concerns the building and civil engineering industries, and the CRINE Report on the offshore oil and gas industries. The basic recommendations in both reports are similar:

- the use of fair contracts;
- the encouragement of cooperation rather than confrontation;
- the development and use of standardized documentation and procedures.

2.12.1 Latham

Sir Michael Latham published his final report on the procurement and contractual arrangements in the UK construction industry in July 1994. The report was commissioned to seek ways in which productivity and efficiency could be improved. The review covered contract conditions and documentation, contracting strategy, payment arrangements, the construction team and settlement of disputes.

The principal matters covered by the very extensive report include:

- Contract conditions: Latham comments that the most widely used forms of contract in the industry reflect arrangements which rarely apply on site, e.g. design complete prior to commencing on site, acceptance by both parties of the architect/engineer as being impartial; he advocates the use of fair contract conditions which reflect the realities of a modern construction project. The report suggests that the New Engineering Contract, issued by the Institution of Civil Engineers, contains a fair balance of risk and recommends that owners should not be allowed to amend standard clauses to gain an unfair advantage.
- Team working: cooperative, non-confrontational working is recommended using integrated teams not only between owner and main contractor but also between main contractor and subcontractor.
- There should be greater use of coordinated contract information.
- Trust accounts: the report recommends the establishment of trust accounts to guarantee interim payments to main contractor and/or subcontractor in the event of the insolvency of a company higher up the payment chain. Funds should be placed in trust accounts at the commencement of each 'stage' of the work.
- Retention bonds: retention bonds, decreasing in value as the work progresses, should be used in place of the common practice of withholding part of the due payment until satisfactory completion.
- Alternative dispute resolution: a mechanism for dispute resolution should be incorporated into the contract conditions.
- A construction contract Bill should be introduced to give statutory backing to newly amended standard forms and to introduce recommendations of the working party on construction liability law.

2.12.2 CRINE

The CRINE Report (Cost Reduction Initiative for the New Era) has been published against a background of low oil prices and rising costs of development of oil and gas discoveries in the North Sea. It is feared that unless costs are reduced significantly within a short time the development of the remaining reserves in the UK sector of the North Sea will be in serious jeopardy. The report makes recommendations that it suggests will have the effect of reducing capital costs by at least 30% thereby improving the economics of present and future fields and extending the life of the UK continental shelf development.

The six major recommendations are as follows:

- to use standard equipment;
- to use functional specifications;
- to use criticality to determine documentation requirements;
- to simplify or clarify contract language and eliminate adversarial clauses;
- to rationalize regulations on certification, production consents, pipeline works authorizations and field development programmes;
- to raise the credibility of quality qualifications.

2.13 INSURANCE

Insurance is a specialist subject and as such is covered in Chapter 8 on insurance and indemnities.

The types of insurance required on a major engineering project are normally very extensive and the scope, sums insured and limits of indemnity have to be carefully assessed, requiring a complex policy statement. A typical engineering construction project is complex, of high value and requires the appointment of more than one major contractor. It is therefore frequently the case that the owner will specify in the contract conditions that he will arrange certain specified insurances.

Such insurances are likely to cover as a minimum the provision of an all risks policy on the permanent works, including material loss or damage to the work undertaken and at the same time protecting the interests of the owner and his contractors. In certain cases, and especially where work is being undertaken in and around existing plant of high value, the owner may also accept responsibility for arranging third party insurance.

2.13.1 Project insurance

An owner's all risks policy will cover the permanent works, materials for incorporation therein and will sometimes extend to temporary works, constructional plant and equipment. The cover will apply on the site of the works, material and equipment in transit and perhaps to off-site work. Either in the same policy, or by a separate insurance, the 'all risks' may be extended to include an indemnity against legal liability for injury to third parties and damage to their property for a specific limit of indemnity.

2.13.2 Contractor's insurance

The extent to which a contractor is responsible for loss or damage to the works and material for incorporation therein will be specified in the conditions of contract.

In addition a contractor will normally be responsible (sometimes without recourse to the owner even where negligent) for loss of or damage to constructional plant and equipment and temporary accommodation.

The contractor is required to undertake by law certain insurances such as employer's liability in relation to his employees and insurance of motor vehicles on the public highway. Other insurance arranged will depend on the contractual obligations and otherwise be based on the philosophy of the contractor's management.

2.13.3 Claims and reporting

It is normal for the contract conditions to require a contractor to report all loss or damage immediately to the owner, there being thereafter a laid down procedure for the reporting of claims and the adjustment of losses depending on which party has arranged the insurance.

BIBLIOGRAPHY

Association of Project Managers (1994) *Body of Knowledge*, Association of Project Managers.
Hackney, J.W. (1995) *Control and Management of Capital Projects*, McGraw-Hill.
Harrison F.L. (1985) *Advanced Project Management*, Gower.
Karger, D.W. (1991) *Strategic Planning and Management*, Dekker.
Kimmons, R.L. (1991) *Project Management Basics*, Dekker.

Estimating

3.1 INTRODUCTION

The ability to forecast cost accurately is important in all industries, but essential where substantial sums are to be committed to a project whose final cost may not be known for a decade or more. A poor estimate can in such circumstances encourage investment in uneconomic or marginal projects while leaving other projects starved of cash, and may cause the failure of a company. Alternatively, opportunities may be missed with the possible risk of a competitor investing in a similar project.

In the engineering industries an owner will seek to estimate demand and therefore the value of a product, balance this against the cost of production, and it is hoped identify an acceptable return on the capital invested. The cost of production will itself comprise estimates of raw materials, the capital and financing costs, and the operational costs involved in processing the raw materials into the product, including subsequent distribution and sale.

This chapter deals with the estimation of the capital cost of production facilities up to the preparation of the most detailed estimate, typically called the definitive estimate. Estimates required for the continuing control of change are covered elsewhere in this book.

3.2 INITIAL ACTIONS

On receipt of a request for an estimate, the estimator must make urgent contact with all the key personnel associated with the project to determine:

- the programme for completing the estimate which should cover provision of drawings, specifications and the like;
- nature of the plant;
- the quality of available information;
- the range of design solutions available;
- an acceptable base case;
- the date for provision of additional information;
- the accuracy and timing of the required estimate(s).

The estimator will need to obtain information regarding:

- the nature of the local environment;
- availability of local services and facilities;
- the position of the plant relative to other constructions and whether such other constructions are 'live' plants;
- relationship of the plant to associated installations;
- whether associated installations provide services to the new plant or are dependent on it.

The estimator will therefore not only consider work within the designated area of the plant, but also the manner in which the facility relates to other installations that may be providing the power, water, fire fighting systems, etc. needed to service the facility. Such systems are, in a petrochemical environment, commonly referred to as utilities. Similarly, consideration is given to the pipelines or other means of delivery, storage or removal of the raw materials and product (commonly referred to as offsites).

Once an understanding is reached as to the nature of the project, consideration is given to the information readily to hand, which can be used directly or in an adjusted form in the estimate. This information may be in house or published, but it will be of little use and positively dangerous if details are not available that place the various costs into context, e.g. location, date of construction, capacity, feedstock, nature of the ground and provision of utilities.

Items frequently undervalued or forgotten from an estimate include:

- utility supply(s);
- extent of access platforms;
- temporary laydown areas for outfitting work prior to erection;
- foundations for heavy lift cranes;
- spares and commissioning spares;
- initial charges or catalysts for the equipment;
- logistical requirements in getting large numbers of men and large items of hardware on to a site.

Other factors could include:

- required skills not locally available;
- local accommodation not adequate to house the workforce;
- bridges not sufficiently strong to carry the transport loads;
- power lines too low or bends in roads too acute to take long transporters.

None of the above can be anticipated from a plant layout, and it is therefore vital that the estimator should inspect the site, the local environment, the main access routes and consider the construction methods to be used to take full account of any such ancillary cost.

Onshore and offshore estimating checklists are available from the Association of Cost Engineers.

3.3 ESTIMATING ACCURACY

An estimate is basically dependent on four factors, each of which will have a significant impact on the accuracy of the figures produced:

- availability of design;
- availability of relevant cost data;
- the skill and experience of the estimator;
- time available to produce the estimate.

When all four of the above are readily available, an accurate estimate should be possible, but in the very early stages of a project, design will not be available and reliance will be placed on historical cost data and experience in the use of such data. Unique projects will require additional cost research to interpolate the available historical data, which further emphasizes the importance of an experienced estimator's ability to interpret the most rudimentary concept into a useful estimate.

It will be necessary at the earliest stages to qualify the estimate regarding any assumptions made which will have a significant impact on the estimate.

As a project develops, the level of detail within the estimate, and therefore the accuracy of the estimate, increases. Not only does the technical definition increase during the duration of the project, but also the number of potential design solutions decreases, thereby reducing the level of design uncertainty and variability.

Although a high degree of accuracy is always to be preferred, there is no merit in attempting to achieve a level of accuracy inappropriate to the information available and inappropriate to the decision that is dependent on the estimate. An estimate produced from minimum information cannot be appreciably improved by attempting to add isolated items of detail, and moreover may give it a perceived credibility that it does not deserve.

The estimator will be required to produce a statement regarding the accuracy of the estimate inclusive of a contingency allowance as discussed later in this chapter. This statement is normally expressed as plus or minus a stated range, sometimes referred to as tolerance limits, which will vary dependent on the type or class of estimate. The limits of the range are the minimum and maximum values of the estimate and are expressed as a percentage of the expected cost. Since there is a limit on lowest possible cost, i.e. zero, but no equivalent upper limit, this 'confidence factor' will often have a positive figure higher than the negative.

The accuracy statement can only be arbitrary, and each estimate should be assessed on its own merits. However, since the judgement of estimating accuracy is primarily based on the quality of information available, including scope, cost data and productivity, an experienced estimator should be able to identify the risk and contingency requirements and forecast a realistic accuracy level.

3.4 STAGES IN ESTIMATE PREPARATION

A project advances by taking measured steps through a minefield of checks (and cheques!) and balances, involving proposals, estimates, approvals, reviews, changes and further approvals. Each approval will be judged against earlier approvals and the owner will look for the reason for any change, especially if such is detrimental to the viability of the project.

Estimates will be required on a project from the date on which it appears as a bright spark in the firmament of an owner's imagination, to the date on which the project is completed and handed over. The number and types of estimates to be undertaken in the duration of the project will depend on its complexity and nature.

The production of an estimate is itself a not inconsiderable expense, and pressure may be brought to reduce the number of estimates or to limit their extent. In such circumstances, there will be a need to balance the cost of preparing an estimate against the risk in not preparing the estimate.

The first stage of a project (the feasibility stage), will begin with the need for a facility being identified. The definition provided to an estimator at this time will be limited to a description of what the facility has to achieve to meet the owner's requirements. The resulting estimate will be at its most approximate, but since early estimates are used solely to decide to commit to a relatively modest additional investment in carrying out further design, the level of accuracy required will only be relative to the investment decision being taken, e.g. whether the project is likely to be viable and hence the design worth developing.

Following completion of the feasibility stage and acceptance of the estimate, a limited amount of design work will be undertaken to develop what may still be sketchy proposals. This is usually referred to as the conceptual design stage and will cover the activities identified in Table 3.1. Conceptual design may be undertaken by the owner or the owner's chosen designer but will always be undertaken by an organization's most experienced personnel. This will be the first attempt to describe and commit to paper the scope of a plant capable of meeting the stipulated requirements, the characteristics and output of the plant and will identify the required processes and major equipment items together with a possible layout.

The cost of detailed design is substantial and can involve large design teams. Time spent in the essential aspects of suggesting and discounting various design alternatives within a particular concept can be costly in both time and hours. It is therefore frequently the case that a further stage is introduced between the conceptual and detailed design stages. This further stage may be used to add further definition to the conceptual design and is commonly termed preliminary design. Alternatively, it may be used to validate some aspects of a well-developed conceptual design or as a first stage of detailed design, termed front end design.

The decision to undertake either preliminary design or front end design will be dependent not only on the status and confidence in the conceptual design but also on the normal procedures of the particular company or project manager.

3.5 ESTIMATE TYPES

As previously stated, the types of estimate are many, they have varying depths of quality and are suited to quite different and distinct purposes. Each estimate is intended to predict the final cost of the design and execution of a project, and each is capable of a degree of sophistication and accuracy appropriate to the stage of the project.

It has been suggested that the early estimates are not carried out by the quantification of components but rather by the application of historical, all-inclusive rates to product criteria.

Estimate content and proportions will of course vary between differing types of project, and will change during the life of a project, reflecting not only the degree of detail available but also the purpose of the estimate. An

Table 3.1 Estimating accuracy

Technical definition	Project stage	Estimate type	Estimate accuracy (%)
Generalized scope definition			
Required capacity of plant			
Plant location			
Preliminary project schedule			−25 to + 50
Preliminary process diagram			
Main statutory requirements			
Equipment list			
Outline engineering specifications			
Preliminary 'block' plot plan			
Offsites and utilities by system and capacity			−15 to + 25
List of major equipment			
Preliminary equipment data sheets			
Preliminary process and instrument diagrams			
Quotations for unfamiliar items			
Detailed scope definition			
Finalized heat and mass balance calculations			
P & I Ds[a] for process and offsites			
Detailed plot plan			
Detailed engineering specifications			
Operator's requirements			−10 to + 15
Local authority requirements			
Project master schedules			
Information on site conditions			
Local availability of labour and material			
Detailed equipment list			
Completed and approved plant layout			
Electrical single line diagrams			
Detailed equipment specifications/data sheets			
Detailed material take-offs			
Firm quotations from potential vendors			
Quotations from potential contractors			
Commissioning and operating information			−5 to + 10
Installation and fabrication specifications			
Amended design and specifications as a result of safety reviews			
Construction subcontract enquiries			
Production design phase continues			

[a] Process and instrumentation diagrams

Key to design stages: Feasibility

Conceptual design

Preliminary front end design

Detail design

Key to estimate types: Order of magnitude

Appropriation

Budget

Definitive

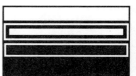

estimate prepared in a formalized manner, with an eye to its continuing use as a source of information and control during the project and as a receiver of information that adds to the in-house knowledge base, will have a number of manifestations. Each manifestation should ideally be a development of that which went before and should accommodate that which is to come.

The exception to the above may be the order of magnitude estimate, which is frequently on a basis different from subsequent estimates. It may be based on the typical cost of a plant capable of processing a stated amount of product per day, which as a set of criteria is incompatible with subsequent estimates, relying as they do on the identification of the labour and materials required to design and construct the plant, rather than the capacity of the plant.

Nevertheless, all subsequent estimates have a hierarchical relationship, each estimate containing a greater degree of detail within a static set of accounts that can be related to the tender enquiry documents and extended into project execution to provide a basis for cost control. Good cost control will keep the last estimate live, with old information being replaced by new for later analysis to provide the high-level data for future order of magnitude estimates. In this way a basis for comparison exists throughout the project life and data are provided for use in following projects.

There are four basic types of estimate that are generally required during the various stages of the project, each of which may be supported by numerous supplementary estimates to assess intermediate change and value options:

- the first estimate during the feasibility stage, variously referred to as a conceptual definition, class IV or order of magnitude estimate;
- the second estimate prior to commitment to the further expense of conceptual design, variously referred to as a outline definition, class III, study or appropriation estimate;
- the third estimate required during the front end design stage prior to commitment to the significant cost of detailed design, variously referred to as a design, class II, preliminary or budget estimate;
- the fourth estimate before commencement of capital expenditure, variously referred to as an execution, class I or definitive estimate.

Estimate titles will depend not only on the industry but also on whether the estimate is produced by an owner or contractor. Each company will have a preferred title, class definition and indicative accuracy ranges, but estimate types are referred to herein as:

- order of magnitude estimate;
- appropriation estimate;
- budget estimate;
- definitive estimate.

In addition, there may be additional intermediate stages, or conversely, a merging into fewer estimate types to suit the individual needs of specific companies. It is therefore advisable to state the probable accuracy, together with the estimate description, whenever such estimates are offered or requested outside the organization.

Table 3.1 demonstrates the phases of the project design during which the technical definition is produced and the degree of accuracy normally expected of the four basic estimate types.

The estimator will concentrate on the critical cost areas, bearing in mind the further effects that such critical items may have on other elements. A typical piping account within a budget estimate for a petrochemical project would be considered critical, since it may equate to 20–25% of total material costs, 40–45% of total labour costs and furthermore will have an impact on pipe supports, foundations, painting, insulation, engineering and project management costs. Piping accounts on an engineering project also have the greatest variability and are the most difficult to estimate.

By contrast, the structural steelwork element of a typical offshore module would be considered important but not critical since although it may constitute a similar proportion of the material and man-hour cost as the previous example, it does not have the same impact on other trades as it is designed early, in a traditional manner, and does not usually change dramatically.

3.5.1 Order of magnitude estimate

Sometimes known as a 'ballpark' or 'seat of the pants' estimate, an order of magnitude estimate is produced for the rapid evaluation of commercial possibilities and economic viability of a project. It is frequently based on gross engineering parameters.

Regardless of title, this type of estimate will be used to gain initial sanction for development funds to continue into the next stage of the project and will contain (relatively) little detail. It will be dependent on historical data and will use broadly based all-inclusive rates. An order of magnitude estimate will typically be considered to have to an accuracy of -25 to +50%.

Since little detailed knowledge of a plant will exist at such an early stage, the estimate will depend on pro-rating information from previous projects that contain similar processes, items, buildings, systems and elements previously designed, built and costed. Adjustments will be made for time, location, changes in market conditions, current design requirements and relative capacity before their use in the estimate.

Cost does not vary in direct proportion to size, and major changes in scope or relative capacity between the current project and the project being used as an estimating base, can be adjusted using a factor sometimes referred to as the 'six tenths' rule, this being a rudimentary ratio by which cost increases relative to size. This rule of thumb states that the amount by which the estimated cost of a desired item should be adjusted relative to a change in size or capacity, can be calculated as the pro-rata change multiplied by six-tenths. On this basis the estimated cost of an item of plant that doubles in size will increase by 60%.

The revised composite rates obtained by such adjustment, or conditioning, provide not only a reasonably accurate and speedy estimate but will also be used to provide a basis for cost control as the design progresses.

Clearly, the level of accuracy achieved from such a 'factored' estimate will depend on the availability of historical data from similar projects in terms of purpose, location and scale, feedstock, market conditions, etc. Data obtained from previous estimates may have originated either from tenders received, or from final costs incurred, and in each case may have to be adjusted further, either because a tendered figure will invariably be less than the final cost, or alternatively, a final cost may include the cost of overcoming problems that may not recur.

If historical data from a previous project which is not a near duplicate of the proposed project are to be used, further research and development will be required to achieve the level of accuracy required, before commitment of development funds.

In the absence of information from a near duplicate plant, the estimator will rely on published or historical information from a number of existing projects to trace the variance in cost of the principal component parts arising from differing duties, capacities, relationships to existing equipment, or the nature of the feedstock.

Historical information is of little use without a knowledge of any qualification that made such data unique in whole or part, i.e. nature of ground conditions, power generation being provided by existing installation, generous or congested plot, etc.

3.5.2 Appropriation estimate

The appropriation estimate is sometimes known as a class III estimate, because it uses technical information developed to a level of definition described as class III.

Although there is still a lack of design detail necessary to achieve even a preliminary quantification of the elements comprising the project, the designers will now have identified the major equipment items and determined their required outputs. This information should be sufficient to allow a more accurate estimate to be made of the cost of the major items of equipment, and will provide an opportunity to make enquiries of potential suppliers regarding key components including budget price. Such key components can comprise a significant proportion of the total capital cost, and the capture of these important costs with a degree of certainty will give added confidence to the estimate.

The remaining elements will still not be defined to a sufficient degree of detail to allow vendors to provide budget prices, or to allow a quantification sufficient to undertake any other than a factorized approach, and in the absence of such better information, the cost of installation of equipment can be estimated by a Lang factor. The Lang factor is named after H.J. Lang, who found that various types of plant had a reasonably consistent relationship between equipment cost and installation cost.

An appropriation estimate will therefore be based substantially on the list of major equipment. Costs will be estimated using in-house data or preliminary budget prices, obtained from vendors for its manufacture and installation, but with most of the remaining construction elements included as a percentage of the value of each item of equipment. Given a good database, such methods can produce a very acceptable result, so long as the nature and basis of the data are fully understood.

An appropriation estimate will commonly be considered to have an accuracy of -15% to +25%.

Vendors will frequently suggest alternative options and consideration should be given to such additional information to optimize costs.

3.5.3 Budget estimate

Sometimes called a class II estimate, the budget estimate will be produced once the conceptual design stage has been completed. It will be used in the final decision to commit to detail design and to provide a basis for the continuing cost control of the project. The budget estimate will typically be considered to have an accuracy of -10% to +15%.

In the early stages of a project, design hours for engineering, procurement, project management and project services will be included in the order of magnitude and appraisal estimates, either by inclusion in a cumulative rate or applied as a percentage. In the later stages of budget and definitive estimates, these same hours will normally be separately estimated by the individual user departments. This will usually be done by the engineer, who will anticipate the number of drawings or requisitions (deliverables) required, or in the case of management, by consideration of the organization required to undertake the work, and the duration of each of the positions identified on the organization chart.

Information available at this stage will allow for approximate quantities to be established and guide prices obtained from potential vendors and contractors. Such quantities, although preliminary, will frequently be available from material take-offs produced from drawings, either manually or electronically.

The cost of construction labour for inclusion in a budget or definitive estimate will usually be calculated by applying man-hour norms to the quantified schedules and attaching all-inclusive labour rates to the resulting totals. The labour rates to be used will include the cost of 'normal' construction plant, supervision, site accommodation, etc. but will exclude major method related construction plant, such as heavy lift cranes and the like.

The material and equipment elements within a budget or definitive estimate will be calculated or refined by the issue of preliminary enquiries for budget prices, once the necessary information is available. In the case of the major equipment items, such enquiries may not be preliminary but rather the request for confirmation of previous estimates, since the vendors will have been asked to provide preliminary information relating to performance, capacity, various aspects of design and estimated price early in the conceptual phase.

3.5.4 Definitive estimate

The final estimate produced immediately following commitment to the major capital expenditure, is the definitive or class I estimate and will typically be considered to be to an accuracy of -5% to +10%, therefore giving little margin for error.

This estimate will typically contain the level of detail used in the execution of projects and the preparation of bids. It will be used in the maintenance of close control over the cost of the work, or for allocation of resources into work packs, job cards and similar project control tools.

The definitive estimate will use a significant amount of the information obtained during the budget estimating phase, but will have the benefit of more detailed design particulars, detailed labour and material breakdowns together with more accurate prices from commitments made, or enquiries issued to vendors and potential contractors.

While the definitive estimate is more accurate regarding quantity, specification and price, since the estimate is being produced at a higher level using specific detail, it is more susceptible to error by omission than is, for instance, an order of magnitude estimate. The order of magnitude estimate will probably have been based on an analysis of the actual final cost of a previous project, a 'top down' estimate. By its very nature, therefore, an order of magnitude estimate includes all costs incurred on the previous project. In contrast, a definitive estimate is a 'bottom up' estimate, built up from information specific to the project and which cannot rely on the ultimate safety net of an all-inclusive rate, to account for all the sundry items that need to be considered when compiling a major definitive estimate. Such sundry items, if not separately addressed, may be included as a percentage add on to the estimate, or alternatively, included within the rates and prices used.

Various on- and offshore checklists are published to assist the estimator and help avoid omissions from the estimate.

3.6 HIERARCHICAL COMPOSITE RATES

All estimates are the result of a quantity being multiplied by a rate, and all rates other than the most fundamental costs are composed of a number of cost elements in various ratios to form a composite rate. Composite rates may reflect detail at various hierarchical levels; the highest level of detail may be the basic costs of labour, plant, materials, etc., while the lowest level could be a single all-inclusive composite rate per unit of production of a completed plant.

Although the use of low level composite rates in an estimate can be both cost-effective and accurate (if properly produced and documented), the use of such rates without an understanding of their original content is dangerous. A full set of composite rates at various levels, without any idea of what the rate represents, when the rate was current, and where they were used, is of little benefit to any company and it is therefore advisable to maintain a record of the basic components of composite rates, including design criteria, date and location, so that rates can be adapted for use in different circumstances.

Unlike the building and civil engineering industries that tend to estimate using composite rates that are inclusive of labour, construction plant and materials, the process engineering industries estimate each of these elements separately.

The building and civil engineering component of an industrial engineering project may initially be produced using methods appropriate to those industries. However, it is not unusual for the estimator to analyse the resulting estimate into its constituent parts for the sake of consistent presentation and to predict more accurately overall labour levels required on site and to anticipate the need to rationalize the labour loading within what is considered to be a manageable total workforce.

The rate for labour used by an estimator will be a form of composite rate usually referred to as an 'all-up' all inclusive rate. An 'all-up rate' can be interpreted as including some or all the costs of employment, provision of accommodation, management costs, construction plant, material costs and the like. It may have been produced from the examination of the total payroll alone, the payroll for a particular discipline, or obtained by dividing the final outturn costs of a project by the total on-site hours expended. An estimator should therefore always qualify his request for information, or his response to a similar request, by stating its inclusions in general terms.

This continuing problem of definition can be further complicated when a rate is stated in terms of direct man-hours, as the estimator must be alert to the further problem of definition, as discussed below.

The message throughout this, and other, chapters is that any rate that is not defined is unusable, and an experienced estimator when offered a rate will always respond with the question 'what does that mean or include?'

3.7 MAN-HOUR NORMS

In the engineering industries, in contrast to the building and civil engineering industries, there is a tendency to estimate labour and material separately. Engineering plants are frequently constructed in remote areas or countries and labour productivity and efficiency is therefore a matter for specific consideration.

Most companies have a set of 'base' man-hour norms; such norms may have been originated by them, or alternatively the company may use as a base a set of published norms. Published norms will usually be at the highest level at which work can be measured, i.e. per metre of a specified pipe or per butt weld, and the estimator may need to compile data on composite norms in the same way as has been discussed for composite rates.

Whatever their origin, norms will reflect a stated set of circumstances and will require adjustment to accord with the levels of efficiency expected of the new plant by the application of an efficiency factor. The efficiency factor to be applied to such basic norms will take account not only of the estimator's experience of his company's recent performance against the established norms, but also the nature and location of the work for which the estimate is being produced.

Efficiency varies from country to country, in regions within a country and towns within regions. It can be influenced by the relative experience of the labour force, site management including planning, trade union influences, staffing traditions and governmental decrees on hours of work, safety and the like. Without any specific experience of his own, an estimator may have to make numerous enquiries to relate the basic norms in his possession, whether in-house norms or published norms, to the particular labour environment, to account for the relative efficiencies of the available labour. In any such enquiries however, care should be exercised so that a full understanding is obtained regarding the information offered.

Productivity and performance should not be confused. Productivity is a measure of the effort required to produce a specified result, whereas performance is the speed by which the result can be achieved. Increased productivity can improve performance, but performance can also be improved by increased labour levels, which in turn may have the effect of reducing productivity and increasing costs.

It should be noted that a productivity factor stated as being more than 1 represents a reduction in productivity while a factor less than 1 is an increase.

Productivity factors should not be applied to any locally obtained budget prices since these would already reflect the local conditions. Comparison of such budget prices with the company standard may, however, provide further means of validating a productivity factor.

Frequently, only 'direct' man-hours are calculated with 'indirect' man-hours being included as part of the 'gang rate' applied to the number of direct man-hours. Direct man-hours are, however, sometimes interpreted as a contractor's own labour as compared to subcontracted labour. Alternatively, indirect man-hours may be understood as all labour other than tradesmen and it is therefore again necessary to define the information offered.

It is useful to use a definition that has an ongoing use throughout the life of the project, by planning and cost control groups. All projects should be consistent to provide a reliable database.

The following definitions of direct and indirect man-hours are suggested:

- Direct labour: crafts or trades that are directly involved with the prefabrication, fabrication, construction, machining, preparation, assembling, erecting, finishing and testing of the work. Direct labour shall include all crafts and trades that directly contribute to the construction of a facility or who are of direct support and without whom the activity could not be achieved.
- Indirect labour: labour of a more general nature that provides support to the direct labour and which is necessary for the orderly completion of the work, but which cannot be assigned to discrete activities and does not directly contribute to the construction of a facility.

Indirect labour may include foremen and supervision, quality control, scaffolding, cleaning, warehouse and transport services, crane drivers, material control, fire guards, safety, dimensional control and construction camp operations.

3.8 ESCALATION, EXCHANGE RATES AND FINANCING CHARGES

On a major project, escalation and provision for fluctuation in exchange rates can be a very significant component of the estimate and it should be made clear in the estimate whether costs include or exclude allowances for escalation.

3.8.1 Escalation

Again, it is advisable to retain within the estimate details of the manner in which escalation was calculated, so that any variances occurring later in the project can be identified and reported.

The estimate should include details of the base date of the estimate, the applicable rates used and escalation allowance and the project programme. The main contract items should be clearly identifiable in a phased escalation programme and anticipated expenditure profiles produced for the most significant activities.

In some cases, owners may predetermine the level of escalation to be assumed, on the basis that the escalation anticipated within their calculations as to the value of the product, will be loosely comparable with escalation in the capital budget.

3.8.2 Currency fluctuations

The need to allow for currency fluctuations depends on the nature of the plant, the location of the plant, the duration of construction and the nature of the owner's business. Some effects of currency fluctuations can be allowed for by the estimator, while others are outwith the requirements of the capital cost estimate.

An example of the latter would be the use of a feedstock priced in a currency that is different from that of the product. This can affect the project in a way that is totally outwith the control of the estimator and the project team.

The effects of currency fluctuation on the capital cost of the plant can, however, be allowed for in a number of different ways:

- Certainty can be purchased by buying forward currencies, in which case the anticipated premium charge of such a purchase should be included in the estimate.
- The owner may be willing to accept the risk of currency fluctuation, in which case the project will agree with the owner a 'project

exchange rate' to be used throughout the life of the project, with the owner taking the risk of currencies varying from this agreed base.

- Owners that are multinational companies with substantial foreign currency holdings may not be concerned with currency fluctuations since they may elect to pay foreign currency accounts out of their foreign currency holdings.

3.8.3 Financing charges

Charges for the financing of a major project throughout the many years of design and construction can be equally important. The estimator will create his cash flow forecast by effectively pricing the project schedule by spreading the expenditure against the periods over which that expenditure is being incurred in line with an appropriate expenditure profile.

3.9 CONTINGENCY ALLOWANCE

A contingency allowance is a sum provided within an estimate for unforeseen items, which experience shows are likely to be required to cover design development, construction change and estimating adjustments within the project scope. Known requirements should be included in the base estimate and not as a component of the contingency allowance.

The contingency allowance should not include for *force majeure*, or for the possibility of specific scope changes. For example, the possibility of requiring piled foundations should be included within the body of the estimate, while the cost of excessively long piles should be accommodated by a specific allowance within the contingency sum. The possibility of the requirement for additional piles in the event of the plant capacity being increased should be excluded.

Similarly, since it is almost certainly the case that general inflation will cause costs to rise, this should be provided for within the estimate under a specific heading; an allowance for unexpectedly high increases would, however, be accommodated within the contingency allowance.

The contingency allowance is not added to the base estimate to reflect the highest possible cost but rather the most probable cost. It will therefore not relate to the extreme range of the accuracy statement referred to in section 3.3 on estimating accuracy.

The amount of contingency to be included in an estimate, although clearly dependent on the information available, is also materially affected by the knowledge of the type of plant being contemplated. An order of magnitude estimate, which is based on the analysis of a recently constructed similar plant, may have an accuracy similar to a budget or definitive estimate, and care should be taken to ensure that any estimate that is based on actual recorded cost is not penalized by the addition of an unnecessarily high contingency allowance.

As stated previously, the contingency allowance can be a very significant account within an estimate and as such deserves to be handled in a manner appropriate to a major cost element. On a major project and where an individual's experience allows, it may be necessary to introduce a probability analysis, or full risk analysis, of the various maximum and minimum contingency figures established, to provide a realistic total allowance.

It is to be anticipated that estimating accuracy will increase in line with improvements in experience, estimating techniques and estimating data. It should always be remembered that improvements that reduce estimating errors or omissions or more accurately predict cost or quantity, should also allow a reduced contingency to be included. Failure to adjust the contingency allowance may result in a high estimate and jeopardize a project.

Management of contingencies is covered in Chapter 5, Part 1 on cost control.

3.10 RISK ANALYSIS AND EVALUATION

Chapter 2 stated that a contingency allowance is a sum provided within an estimate for unforeseen items, but not for items such as *force majeure* or other extreme happenings that are totally outwith the control of any management. It also identified that on a major project it may be necessary to undertake a full risk analysis and thereafter manage the risk through all the stages of the project. The analysis, identification and management of project risk is discussed in Chapter 2.

3.11 ESTIMATE CONTENT

An estimate prepared for the design and construction of a capital project will contain many sections, usually called accounts, which may be summarized under headings. On a process engineering project these would be typically as follows:

			%
1.	Design, procurement, project services		**10.0**
	of which		
	•	engineering design	6.0
	•	procurement	1.5
	•	project services	2.5
2.	Construction labour/supervision		**30.0**
	of which		
	•	civil	8.0
	•	piping	14.0
	•	electrical/instrumentation	6.0
	•	paint/insulation	1.0
	•	other	1.0

			%	
3.	Equipment		**20.0**	
	of which			
	•	vessels/exchangers		10.0
	•	rotating equipment		6.0
	•	other		4.0
4.	Materials		**20.0**	
	of which			
	•	civil		4.0
	•	piping		8.0
	•	electrical/instrumentation		4.0
	•	paint/insulation		1.5
	•	other		2.5
5.	Site/construction overheads		**5**	
6.	Escalation		**5**	
7.	Contingencies		**10**	
	Total		**100**	

3.12 INTERFACES

While producing an estimate the estimator will constantly seek to question the information provided to ensure that he has an understanding of the full scope of the work and to satisfy himself that the information provided by others also reflects the full scope as he understands it. Examples of queries raised at this interface will include the following.

3.12.1 Process

Review equipment list with the process engineers to ensure that equipment not related to the actual process is included, i.e. service cranes within a compressor building, etc.

3.12.2 Procurement

Establish with the procurement department that enquiries for check prices include all anticipated elements of cost, i.e.:

* delivery and documentation costs;
* construction and commissioning spares;
* full painting specification;
* suppliers' site engineers;
* special tests.

3.12.3 Piping

Establish confidence level of material take-offs produced by piping engineers and agree any adjustment necessary. Check with the piping engineers an inclusion for cutting and waste, which is an allowance for materials that have to be purchased in excess of the installed quantities, including allowances for short lengths of pipe not used, damaged, lost, etc.

3.12.4 Construction

Construction possibly carries the highest degree of risk to an estimator and care and diligence must be used when estimating construction costs. The opinion of the construction department should be sought for the following:

- selection of appropriate construction norms;
- productivity in relation to site location, degree of pre-assembly, establishment location, construction time, e.g. predominantly winter or summer, working elevations, proximity to other major construction projects, working within the confines of live plants, requirement for working permits, etc.;
- labour availability;
- site working agreement including normal working week, requirement for shift working and productivity payments, radius allowance reflecting local labour market, etc.

3.13 MONITORING CHANGE

Once the conceptual design is completed and work on the budget estimate begins, it will be necessary to ensure that changes or potential changes to the estimate are identified. Clearly with information at a preliminary stage, and with a number of assumptions and approximations being made, it is sometimes difficult to keep up with the possible changes that can occur from any given or assumed base.

It is therefore necessary to identify the key documents on which all other aspects of design rely and to establish a specified revision as the base and to monitor change against that base.

There are five documents that are fundamental to an engineering design, control of which will identify all significant change:

- the basis of design or design specification;
- the process/utility flow sheets and heat and mass balance calculations;
- the piping and instrument diagrams;
- the equipment layouts/plot plans;
- the electrical single line diagrams.

Any change identified to these documents should be investigated by the estimator and the cost ramifications assessed by using his own knowledge or that of the design engineers, following which the necessary reports should be raised or adjustments made to the estimate.

3.14 CONTROL BY ESTIMATE

Costs require control at all times but control becomes progressively more important as the rate of expenditure increases. Expenditure will start to accelerate once the conceptual design is completed and cost information must be made available to set targets, record progress and expenditure, anticipate corrective action and forecast final cost.

Cost that is being progressively expended over a significant period cannot be controlled against a single high value entry. Although the definitive estimate may contain the required detail it will not be available at an early stage and if, for example, the budget estimate contains a single figure for the bulk purchase of carbon steel pipework, this will not allow effective control of the individual purchase orders to begin.

The detail contained in a budget estimate may therefore require expanding or refining before it can be used as an effective working tool. If information is not available to produce an accurate breakdown the estimator must make his best assessment since even an arbitrary figure provides a basis for comparison and control.

3.15 CONTINUED USE OF ESTIMATE

An estimate should not be put aside once the total has been approved. The estimate should instead be part of a dynamic ongoing basis of control of all actual expenditure or anticipated expenditure in order to best anticipate and control the final costs. This control is to be exercised by cost sampling and cost control techniques covered elsewhere in this book.

3.16 CODING AND CONTROL

In order to establish a system of control during the life of a project, means of comparison between projects, a formalized approach to the production of historical data, and a good workable coding system should be established which recognizes the various estimating and cost control requirements.

Codes of accounts are further discussed elsewhere in Chapter 5 on project controls, but their relevance to estimating is that they formalize reports into a structure that will facilitate future estimates. Regrettably, each user body tends to have its own code of accounts and unless contractors run parallel systems, data captured on one project are not as easily translated into a new project, and computer usage is therefore restricted and less efficient.

A good coding system will provide the three or four top levels of detail and make allowances for the further detail that would be required by, for example, an enquiry based on unit rates. It should also reflect a job execution approach since this provides means for ongoing monitoring and control.

3.17 ESTIMATE PRESENTATION

The detail provided with an estimate will depend on the requirements of the recipient, but whether or not the detail is to be submitted, the estimate should be fully documented to provide a basis for estimate revisions and reallocations and to provide a basis for feedback analysis. Estimates produced in a consistent manner will assist in the establishment of a basis for comparative analysis and in-house estimating.

It is, therefore, preferable to use standardized forms, each dated, and containing clearly and concisely, all calculations, assumptions and estimate basis. Of course, the use of a computer in storing and analysing such data is of immense benefit.

BIBLIOGRAPHY

Association of Cost Engineers (1991) *Estimating Checklist for Capital Projects*, E & F N Spon.

Association of Cost Engineers (1981) *Offshore Checklist*, E & F N Spon.

Bentley, J.I.W. (1988) *Construction Tendering and Estimating,* E & F N Spon.

Collier, K. (1988) *Fundamentals of Construction Estimating and Cost Accounting*, Prentice-Hall.

Holland, D.M. (1985) *Measuring Profitability and Capital Costs*, Lexington.

Institute of Chemical Engineers and Association of Cost Estimators (1989) *Guide to Capital Cost Estimating*, Institute of Chemical Engineers.

Kharbanda, O.P. and Stallworthy, E. (1991) *Capital Cost Estimating for the Process Industry*, Butterworths.

Smith, N.J. (1995) *Project Cost Estimating*, Thomas Telford.

Spain (1991) *Budget Estimating Handbook*, E & F N Spon.

Taylor, T. (1987) *Electronics Industry Cost Estimating Data*, Wiley.

Value
Management

Sheildhall sewage works, Glasgow, Scotland for Strathclyde Regional Council.

4.1 INTRODUCTION

Value management (VM), value engineering (VE) and value analysis (VA), are terms that are being used more and more frequently; but what do they mean? Most design organizations have a concern for value and have processes and design reviews to address owner requirements and the cost of any proposed scheme. Capital cost is certainly one element of value but general experience indicates that the cheapest is not always the best and a broader understanding of value is needed.

It is increasingly recognized that not only do design organizations need to provide assets that meet the owner's specification and are built to time and within budget but that the potential project maximizes the benefit to the owner's business. At the earliest stages of project thinking the owner specification is an expression of a number of perceived and/or actual 'needs'. As design proceeds and the cost of meeting those needs emerges it is often necessary to recycle the design to provide a scheme that is more acceptable to the owner. In the same way, as design proceeds, additional opportunities can be generated that could bring additional benefits to the owner's business.

It is a key role of the design organization to enter this debate of value and need at the earliest stages of design. Experience shows that 'loose discussion' and 'asking the owner what is wanted' is unlikely to be enough, and design organizations need to be able to get close to the owner, understand the needs and provide solutions which not only give good capital value but also enhance the business.

It is this presentation of need and cost in a different form that VM techniques address. VM and associated techniques should not be seen as displacing normal design reviews but as providing a different perspective to the owner which is not apparent from the normal engineering specification and cost estimate.

4.2 DEFINITION OF VALUE

Value can be described as the relationship between satisfaction of need and the cost of resources required to achieve that satisfaction. This can be expressed as:

$$\text{value} = \frac{\text{satisfaction of need}}{\text{cost of resources}}$$

From this relationship it is clear that value can be enhanced by improving the level of satisfaction as well as reducing the resources needed. In some cases the level of satisfaction might be improved to such a degree that an increase in resources is justified. In most cases capital cost is an important resource but operating and lifecycle costs and availability of skills could figure in the equation.

It is worth recognizing that different interests in a project have different needs; for instance a factory manager could be looking for ease of maintenance while the sales manager might value a quicker order lead time. VM techniques seek to provide mechanisms by which these differing interests and needs can be brought together.

4.3 VM DEFINITIONS

Although the terms value management, value analysis and value engineering are widely used, there is not total agreement on definitions. The European Union is working on an overall approach to the understanding, practice and training for value management within the EU and clarification of terms is expected.

However, it is generally accepted that VM covers a range of value improving practices, particularly those used for early conceptual thinking and the 'softer' front end issues of a project, while VA and VE are more usually used to cover studies of the traditional 'harder' issues at the later stages of design. Commonly, the term VE is used for studies at the project cost level with VA used to describe more detailed component analysis. However, some organizations transpose these terms.

It is possible to consider four broad levels of review as shown in Figure 4.1.

The cost of change at the later stages of design is also greater as design proceeds. Although obvious, this point is often forgotten and expectations can be created that VE/VA studies at the later stages of design will deliver substantial cost–benefit opportunities. This is unlikely as major changes

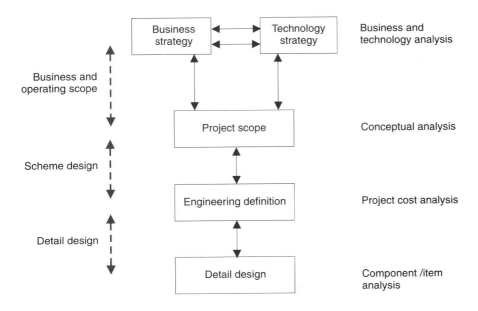

Figure 4.1 Levels of review

cannot usually be tolerated because the delay and cost of making changes is unacceptable. This often results in only minor changes of a cost cutting nature being made, which should have been identified by normal design/cost reviews. Thus the distinctive contribution that VM can make is not realized, and at worst the VM process is devalued.

The reviews can be described as below:

- A technology/business review confirms the business strategy and establishes a broad understanding of the technology and overall operating and other requirements.
- A conceptual review establishes or confirms the overall project scope and, if orders of cost are available, confirms that the broad design and requirements are considered 'good value'.
- A project cost review is a more detailed examination of a proposed scheme carried out when the specific engineering elements have been specified and can be looked at to confirm that the design is optimal in meeting the requirements defined at the concept stage.
- A plant item or component review considers the design of specific items.

4.4 TIMING OF STUDIES

The opportunity to make savings or other changes diminishes as the project approaches detail design and financial authorization is given for purchase of equipment and other engineering works – see Figure 4.2.

4.5 VALUE MANAGEMENT METHODOLOGY AND JOB PLAN

In general VM embraces three elements which together provide the uniqueness of the approach. These are:

1. team based working;
2. a structured 'job plan';
3. use of the principle of 'functionality' or 'need'.

The first two are not unique and are characteristics of many problem solving techniques. The latter is the distinguishing feature of most forms of VM/VE/VA techniques. 'Functionality' is used to describe what the component, system, plant or project is trying to achieve. For example, the 'function' of a door might be to provide privacy, provide security, create the right image or a combination of all three. This distinction provided by using 'functional' descriptions is not apparent from the engineering description 'door'. Describing components in this way enables costs of items fulfilling the same function to be added together, thus allowing the owner to understand the overall cost of that function. Taking the previous example, it might be that if an element of the door is primarily for image then other elements that are there for the same reasons can also be included

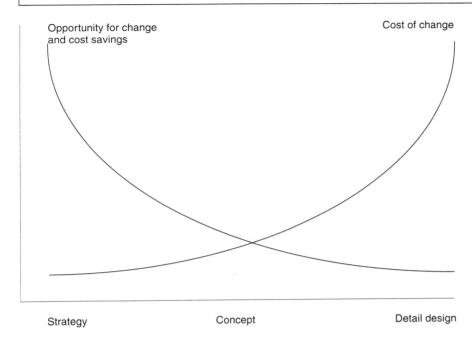

Figure 4.2 Opportunity/cost of change relationship within project timetable

within the functional description. A better understanding of the overall costs and options for providing the image function can be assessed – including that for the managing director's office! 'Functional' descriptions also differ from conventional engineering descriptions in that many items of equipment fulfil more than one function without this distinction being apparent from the cost estimate.

The principle of functionality can also be used to help develop the scope of a proposed project. A successful chemical plant project will not just have the main process function of 'react feedstocks' but will probably have other process functions such as 'recover materials', 'remove impurities', 'achieve product form', etc. There will also be functions such as 'ensure operability', 'make safe', 'protect the environment', 'provide for the future', 'improve customer service', etc. These are all typical functions for a process plant, the costs of which are not apparent from a normal engineering specification and cost estimate. It is these other functions that determine whether an asset will be 'world class' or ahead of the competition and not just a copy of that being used by others. It is critical, therefore, that these functions are properly defined and explored.

It is normal to present functional descriptions in the form of a FAST – functional analysis system technique – diagram. A feature of this type of diagram is that the hierarchy of needs/functions enables costs and opportunities to be understood at different levels. An example of a FAST diagram

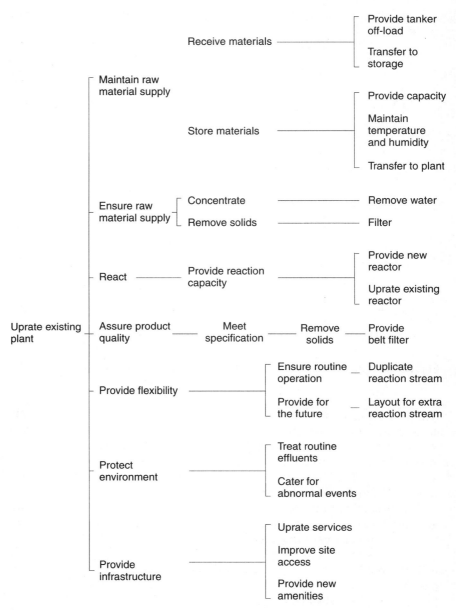

Figure 4.3 Typical FAST diagram

for a chemical plant is shown in Figure 4.3. From this it can be seen that 'assure product quality' could be considered at different levels, i.e. change the product specification, change the process technology to give a more effective process, change the equipment type or change the detail design of the selected equipment. All of these present opportunities with different levels of ease of implementation and trade off. This form of presentation gives an enhanced view of the project that is not apparent from a normal engineering estimate.

A feature of the FAST diagram is that it is developed from the left-hand side by asking 'how?' and from the right-hand side by asking 'why?' It is this questioning that enables the fundamental purpose to be established. For instance, asking 'why a distillation column?' could give a variety of answers including: 'remove impurities', 'recover material' or 'concentrate the product'. These different descriptions give a quite different perspective of the need for a distillation column and the possible options, e.g. different process, equipment, operating (get someone else to do it) or commercial. A further feature of the FAST diagram which leads to greater clarity is the use of 'verb-noun' descriptions of functions. For instance 'increase product quality', might probably require a quite different solution from 'maintain product quality'.

4.5.1 The job plan

The job plan helps to work in an ordered and systematic way. The main phases of the job plan that are commonly used during design development are as follows.

Information phase

The information phase involves defining the project, gathering the available information, establishing constraints and carrying out a function analysis to describe 'What is it we want the project to do?' rather than 'What are the physical components of the project?' It is at the start of this phase that the level of study and potential team members should be defined.

Creative phase

During the creative phase the team uses the functional descriptions, normally presented in the form of a 'function' or 'FAST diagram' to explore alternative ways for accomplishing the desired functions. 'Brainstorming' is usually used to generate ideas.

Evaluation and analytical phase

The ideas generated during the previous phase are screened and evaluated by the team during the evaluation and analytical phase. Normally, some form of decision evaluation technique would be used to help during this phase.

Development and recommendation phase

During the development and recommendation phase the team reviews the selected ideas and agrees on the proposals to be adopted or requiring further investigation.

Presentation phase

The presentation phase involves the presentation of the agreed design proposal including any outstanding developments.

Implementation phase

It is important to see that the recommendations are developed and progressed during the implementation phase. Often, a lot of energy is put into the first five phases over a fairly short period but as implementation is over a longer period and involves people not involved in the initial review, momentum can be lost.

The job plan is not meant to be prescriptive but is there to help a team to work systematically through the design phase. Judgement is needed to structure the job plan to suit the level of study and complexity of the project.

4.5.2 The study team and resources

There is a view that studies interrupt the design process. The opposite is the case. Properly set up studies enable the interested parties to come together and conduct a comprehensive and orderly review of the project, particularly at the early stages of design. In many cases the study can take the place of normal design reviews and team meetings. Reviews present the opportunity for people not normally engaged in the design process to be involved. It is only the owner and interested parties that represent the owner's various interests that can change the scope of the project, the engineering team can make recommendations, and then only if they understand the wider aspects of the project.

As for most team based activities, the size of the study is important. Too many or too few participants results in the study being unmanageable or there not being sufficient interaction of ideas. A team of five to eight people is ideal and selecting the team often gives a good insight into who is really important in the project decision process.

The formal team based stages of a study, i.e. construction of the function diagram, speculation and evaluation, can normally be carried out over a half to two days for each phase with breaks to summarize and generate more data. Thus, over a period of one to two weeks, most small to medium-sized projects can be reviewed. In some cases a very detailed analysis of a large scheme down to the component level will take longer, but note needs to be made of previous comments on the opportunity to make significant changes.

4.6 CONCLUSION

Although the value management techniques that have been described are based on well-established and proven techniques there is the need to recognize that the experience and track record of the 'study leader' is critical to success. For any individual to become proficient and suitably experienced it is recommended that study leaders should be carrying out approximately one study a month. There is the temptation to believe that engineers and designers who have attended a VM training course are able to lead VM reviews. The experience needed to underwrite a sound theoretical understanding of the techniques can only – as in any activity – be gained through practice and the personal learning that takes place. The techniques need to be used flexibly and with imagination: a too mechanistic application can sometimes miss the issue and the study leader needs to be aware when to adapt the technique to draw out the real issues. This requires the study leader to be able to draw on other techniques and methodologies to support development of solutions to the issues raised by the formal 'function' analysis.

Project
Services

Tiffany Offshore Oil Production Platform,
160 miles North East of Aberdeen, Scotland.
Operator AGIP (U.K.) Ltd.

Showing in foreground,
Safe Britannia (temporary)
accommodation vessel.

5.1 INTRODUCTION

Being in control implies a knowledge not only of the current position and ultimate target, but also of the route to be taken to achieve that target. The target will be to complete the work within time and budget, while the current position will be described in terms of progress achieved and cost or time expended. By monitoring past performance on a project it is possible to forecast future results. From these forecasts it may be evident that the achievement of the target is not possible without making changes in the project strategy. It should be noted that these changes may have an effect on both time and cost; due consideration needs to be taken of both. It is also necessary when reviewing past performance to understand the factors that influenced that performance; a judgement has to be taken when forecasting future results as to whether these factors will still apply or indeed whether new factors may come into play and influence the future outcome.

Control is an ongoing process which seeks to:

- monitor performance;
- understand those factors which influence performance;
- suggest corrective action where performance is adversely affected;
- forecast future results;
- suggest changes in strategy where the forecast result does not meet project objectives;
- provide a basis for continuing performance improvements.

5.1.1 Project services

The various factors which influence a project affect both time and cost and it is therefore necessary to ensure that records of progress and cost are compatible in order to assist in the identification of future trends. Many companies recognize this strong link between cost and time by placing responsibility for both with a single group frequently called project services.

Since the receipt and issue of documentation and free issue materials is governed by the programme the group may also include both document control and material control. Project services usually covers all those matters which relate to the direct or indirect control of cost, time, materials and documentation, throughout the life of a project.

5.1.2 Project controls

Project controls comprise a suite of procedures and reporting structures providing appropriate information, at the right time and in as much detail as will allow the project team to make informed decisions. Project controls are not an end in themselves but are effective tools for use by the project team.

Procedures and controls must be seen by all users and contributors as assisting with the management of a project and not as obstacles to be avoided or overcome. If the reporting structure and suite of procedures is relatively easy to understand and efficient in use it will be acceptable and complied with by all contributors. Conversely, if the level of detail requested and the time needed to provide it is so excessive that it is viewed as a drain on time and effort, it will meet with resistance, leading to poor quality data and unreliable/meaningless results.

Part 1 Cost control

5.2 OBJECTIVES

The Association of Cost Engineers defines cost control as 'The process of controlling all cost factors connected with a job so that production facilities for a defined duty are achieved at an economical cost within the amount of money appropriated'. In order to achieve this, cost control activities on any project should encompass the following objectives:

- identify potential cost trouble spots in time for corrective or cost-minimizing action to be taken, i.e. detect potential budget overruns before they become a reality;
- ensure that those who are spending the project funds are aware of the budget for their area of responsibility and how their expenditure performance compares to that budget;
- create a cost-conscious environment so that all within the project team are cost conscious and aware of how their activities impact on the project cost;
- minimize project costs by looking at all activities from a cost reduction point of view.

To meet these objectives it is necessary to:

- have a realistic financial yardstick (the control estimate);
- break down the control estimate into controllable packages of work utilizing code of accounts and/or work breakdown structures as appropriate;
- generate accurate and timely cost forecasts for these packages of work;
- compare these forecasts with the yardstick to identify potential problems;
- take positive action to minimize these potential problems.

The essence of any type of control is the establishment of an agreed defined base, the monitoring of results against that base, the identification of adverse trends, the proposing and implementing of corrective measures and the monitoring of such measures to ensure that they have the desired effect.

Clearly the identification of adverse trends must be in sufficient time to allow corrective actions to be taken, and therefore cost control is an on-going continuous activity rather than an occasional event.

It should be noted that cost control cannot be viewed in isolation. As stated in the introduction to this chapter, cost and time are interrelated where an overrun on time often results in an overrun on costs. It is necessary therefore to consider progress, value of work done, productivity factors, etc. when deriving a cost forecast. This is explained in greater detail later in this chapter.

Cost control does not necessarily mean that a job is done in the cheapest possible way, but it should ensure that the project will be completed at an economic cost within the agreed estimate. It should also ensure that there is no overdesign or uneconomic design.

5.2.1 Cost control terminology

It is necessary to understand some of the terminology frequently used in cost control and by cost engineers. Confusion often exists as to the difference between commitment, accrual and expenditure, detailed records of which are maintained by cost engineers for comparison with a control estimate. These terms do tend to vary in minor ways across the industry, but a common definition is as outlined below.

Commitments

Commitments against a budget within the control estimate will normally comprise the following:

- value of orders placed;
- faxes or letters of intent issued;
- estimated values of orders with no fixed value;
- allowance for escalation (which will include cost of freight, carriage, duty and taxes).

Accrual

The accrual is the difference between the invoiced value and the actual value of work performed, be it design engineering or construction installation work, at a given point in time. While to some extent an approximation, the invoiced value plus the accrual is normally a reasonable assessment of the value of work done. It is therefore the major tool used for forecasting as commitments are recorded in advance of work actually being accomplished and expenditure lags behind.

Expenditure

Expenditure is a record of 'cash out the door'. Due to the time lag between the work being accomplished and the invoice for that work actually being paid it is of little use to the cost engineer as a forecasting tool. However, it is of obvious importance to the owner's accountants who require to know the magnitude of project funds which will be required on a monthly basis to meet invoice payments. As part of his responsibilities the cost engineer is often required to supply cash flow forecasts on a regular basis during the life of the project.

5.3 THE CONTROL ESTIMATE

The basis for controlling costs is the control estimate. The control estimate is developed by taking the best estimate available at project commencement, in most cases the definitive estimate (see Chapter 3), and sorting it into related categories which reflect the way in which the project is to be executed. The whole of the cost control philosophy relies on the control estimate in that all forecast costs derived on the project are compared with the appropriate estimate allowances contained within this document.

In order to ensure that these comparisons are on a like for like basis the control estimate should be broken down in a manner which will be compatible with the way later information will be available. The control estimate should be suitably qualified to indicate to users the level of accuracy which can be expected and access must be available to the full record of information used and assumptions made in its preparation.

Prior to the control estimate being in place it is of course still necessary to control costs accruing against the project. These costs will primarily consist of feasibility and front end engineering activities and must be taken into account when compiling the control estimate. Once the control estimate is agreed, all matters which may have an impact on final cost will be under constant review including changes, predictions of cost, time, resources and contingency allowances required to complete the project.

Further details on estimating are given in Chapter 3.

5.3.1 Changes to the control estimate

Perhaps the most important aspect of cost control in the engineering office or on the construction site is the control of change. The prompt identification of changes to the scope of work and the speedy conversion of these changes into time and monetary terms, where applicable, is of vital importance in the control of a project. Without these, control and cost forecasting systems cannot reflect the true scope of work.

As scope changes are identified, estimated and authorized they must be added to the project control estimate ensuring that the current control estimate always reflects the agreed current scope of the project. The original control estimate was completed at a specific point in time and reflected the project scope as it stood at that point in time. By the addition of scope changes the estimate is kept alive and up to date, always reflecting the current project scope.

Changes to design are often unavoidable due to the nature of plant design. It is, however, essential both to a successful design operation and to effective cost control that design changes are controlled and reduced during the detailed design phase, resisted during the production design phase and accepted only in exceptional circumstances during the construction phase.

Changes must therefore be identified early and brought to the attention of the project manager and owner speedily in order not to disrupt the design effort.

All design documents emanate from a few key documents which formed the basis of the estimate for the project. It is not possible to significantly change the design without changing all or one of these key documents. Thus, if changes to key documents are controlled, changes to the design can be controlled.

It is necessary when considering changes to differentiate between agreed changes and non-agreed changes when reporting. An agreed change is one which is accepted by both owner and contractor as being necessary to the project and for the contractor to have an entitlement under the contract. Such changes would be added to the original control estimate and incorporated into the current control estimate and project schedule. Therefore, assuming efficiency is currently 100%, forecast final cost would be current expenditure divided by the progress to date.

Non-agreed changes may include costs for which the contractor has no right to reimbursement under the contract and 'contractual' changes for which agreement is awaited. The original control estimate should not be amended for such changes and they should not be incorporated into the current control estimate or project schedule. Cost and time has or will be expended on the 'change' and this must be considered when assessing the forecast final cost.

The differentiation is important when reports are viewed by the owner and the contractor. These non-agreed changes, while in many cases being rejected by the owner as justification for additional reimbursement, do highlight factors which may have adversely affected the contractor's performance. These factors therefore may well form the basis of future claims to be lodged by the contractor against the owner.

It is usual to incorporate only agreed changes into the current control estimate and project schedule. In order to ensure that cost and planning are always on the same basis it is important that cooperation between contractor and owner is such so as to provide a rapid response against change requests.

5.4 COST SEGREGATION

The essence of any cost control system is to break down the project into packages of work that can be controlled and managed.

A project cannot be controlled by simply keeping track of the total expenditure accruing on it. After all, using this method, the project team would only realize they had a problem when the total project funds ran out.

If the cost engineer is to identify potential problem areas and adverse trends it is necessary for a project to be broken down into a series of budgets which can be individually reported against. These budgets should reflect both the cost elements of the project, i.e. material and equipment purchases, contractor costs, design costs, etc., and the way in which the project is to be executed, thereby making it possible to allocate budget responsibility to the appropriate members of the project team.

These two differing requirements are satisfied by the adoption of two interrelated reporting structures commonly referred to as the work breakdown structure (WBS) and the code of accounts. The interaction of these structures allows the reporting and control of costs by discipline or by area or by discipline within an area, e.g. total cost of labour, total cost of all requirements in an area and total cost of labour in an area.

When both types of structure are used with a complementary coding system the result is a flexible reporting matrix capable of producing substantial detail if required, while summarizing at various levels and in various formats to suit the requirements of the recipient.

5.4.1 Work breakdown structure

A project requires that time and cost are controlled and hence a hierarchical structure is required which is relatable to these two essential bases of control. As previously stated, control not only involves monitoring actual costs but also the comparison of such costs against a fixed base and the timely identification of adverse trends and recommended corrective actions.

The selected structure is commonly referred to as a work breakdown structure (WBS) and covers the total project with each layer of the hierarchy having successively greater degrees of detail (see Figure 5.1). It will run through owner, contractor and subcontractor activities and will link defined packages of work, their estimated and actual cost and their estimated and actual schedule.

Such packages of work at whatever level in the WBS may be referred to as work packages, sometimes as CTR packages (cost time resource) or ACP (activity control packages) or other titles; but they are fundamentally similar in concept.

On most projects the person who most influences design progress and cost is the design engineer at 'the work face' or his immediate superior, who may be supervising a number of engineers designing a particular section of the work. On a construction site the supervisors may have the greatest influence on progress and cost.

The person who could have the biggest effect on a portion of the work should also have responsibility for that portion of the budget in terms of both time and cost. Most problems start at the detail level and most solutions can be found there and the control structure should therefore not only reach this level of detail but should produce information, such as progress to date, cost to date, etc., which will assist the responsible persons in successfully meeting their responsibilities.

A typical work package will contain the following information which serves to define the budget against which control is to be exercised:

- scope of the work package;
- name of the responsible engineer;
- planning network references and timing data;
- resource man-hours and cost summaries;
- estimated cost of bulk materials;
- estimated cost of plant and equipment.

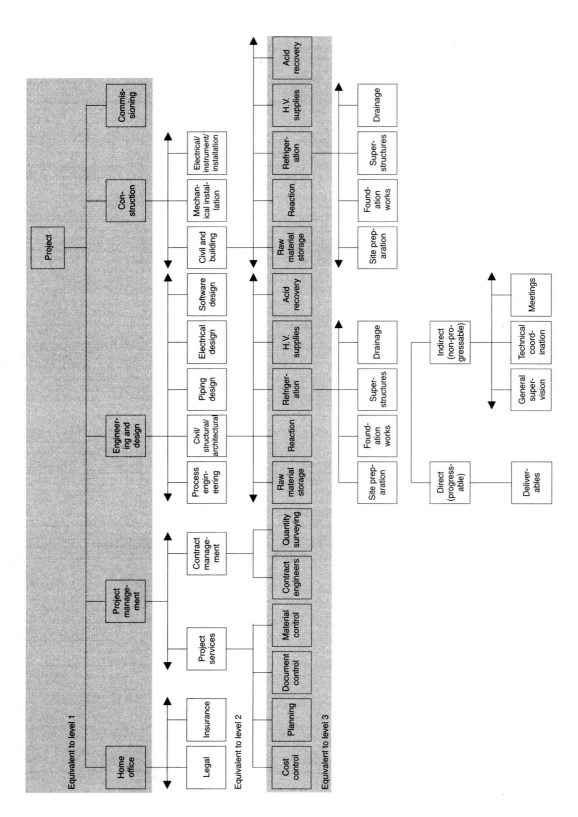

Figure 5.1 Example of a work breakdown structure (showing typical structures)

Work packages should not be so large as to make the identification of problems unduly difficult, nor should they be so small that difficulty is experienced in administering a large number of packages with their various interfaces. Also bearing in mind that the most accurate measure of progress is when an activity is 100% complete and as adverse trends need to be identified early, work packages should not be of an excessive duration.

Consideration should be given to the level at which control is to be exercised. It may for instance be considered necessary to monitor progress on individual drawings only to monitor man-hour expenditure against progress on a group of drawings.

In deciding what level of control is most appropriate the following criteria should be considered:

- accuracy of base information at the lowest level of detail;
- amount of time required to maintain a highly detailed system;
- potential for error increases in proportion to detail.

It is not normally recommended that man-hours expended per document is recorded. Such records produce not only excessive detail but also widely varying results due to the comparison of precise actuals with an estimate at its least precise level.

For a WBS to be effective it must be structured in a manner which will allow information to be 'rolled up' to each successive level of the hierarchy; the top levels will therefore be owner 'owned' and the lower levels contractor or subcontractor 'owned'. The owner will have to define his requirements to the contractor, thereby allowing the aggregation of results at the interface. The contractor will have to ensure a similar arrangement with his subcontractors.

5.4.2 Code of accounts

The code of accounts is sometimes referred to as a cost breakdown structure, because it reflects cost categories, and is sometimes referred to as an organizational breakdown structure (OBS), because it reflects to an extent a company's organizational structure.

Figure 5.2 is based on the standard code of accounts published by the Association of Cost Engineers which defines a code of accounts as 'A logical labelling system utilizing a series of numbers and letters to permit the costs of like or related items to be collected together'. This labelling system provides the basis for reports on expenditure, commitments and predictions of costs to the completion of the project. The primary reasons for imposing a code of account system can be summarized as follows:

- to aggregate data logically;
- to impose a consistent approach to data collection;
- to assist in the control of costs;
- to provide cost data for future projects.

Coding systems vary across the construction industry but normally consist of a hierarchical system making reporting at various levels of detail possible. Thus they ensure that detailed information can be extracted from the

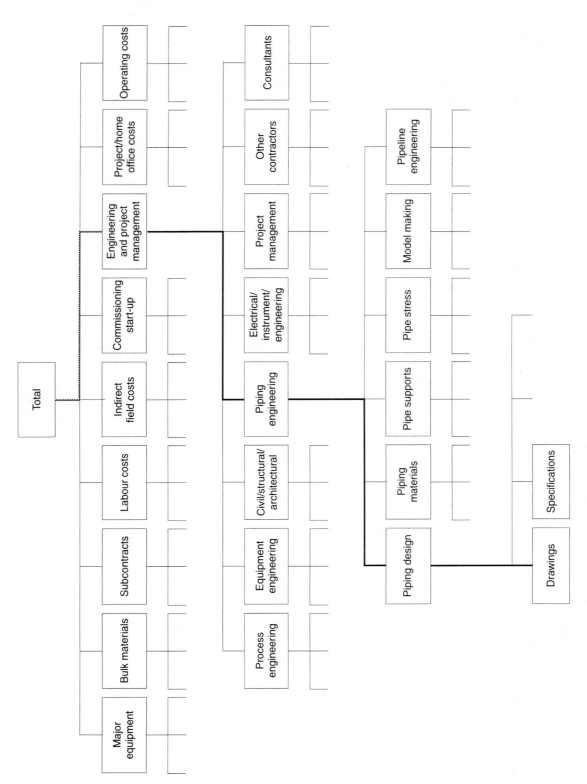

Figure 5.2 Example of code of account

system to assist with future project estimates while supplying high level cost information for cost control purposes.

While providing detailed information on project costs it is also important that the coding system should be as brief as possible and yet easy to use and understand. The cost code may use a series of four alpha numerics to identify categories of cost. Using Figure 5.2 and cost code 7421 as an example, cost code 7*** may designate engineering and project management, 74** piping, 742* piping design with the fourth digit designating deliverables, e.g. drawings, specifications, etc.

It is essential to produce an efficient coding system early in the project which reflects the levels of detail required since once the project is running it will become increasingly difficult to produce detailed cost data below that allowed by the original coding system.

The code of account should allow its use on many different projects. This will not only facilitate accuracy in cost allocation and reporting but will over a period of time assist the estimators by providing compatible estimating data.

5.5 APPROVAL OF FUNDS

Prior to incurring expenditure on a project it is necessary to obtain authority from the owner to spend money. It is normal with large projects for this authority, often termed sanction, to be able to be granted only by the owner's board of directors.

Procedures to be followed when applying for sanction of funds vary between companies; however, it is normal for a justification document for the project to be prepared as a basis for the evaluation to proceed or abandon. It is likely that a specific project will be competing against numerous other projects for limited available resources, particularly finances, and for each project to be assessed on its merits by the board of directors, or other high level of authority, in an effort to ensure that the company's resources are invested in the projects which will give it maximum future benefit.

The justification document will contain all the key information on the project to aid this decision making process. Key information to be included will normally comprise the following:

1. project description;
2. project objectives;
3. technical factors:
   - technology to be used;
   - project location;
   - size of the facility;
   - raw material requirements;
   - product output;
4. commercial factors:
   - market trends;
   - anticipated benefits (e.g. product sales);

- anticipated actions of competitors;
5. financial evaluation (capital outlay versus anticipated return on investment):
  - estimate of project cost;
  - project schedule;
  - project execution strategy;
  - assessment of risks.

As covered in Chapter 3 on estimating, projects move through a series of phases from feasibility studies through to commissioning and handover. It is normal for the sanctioning of project funds also to be phased in a similar fashion. This ensures limited funds are made available during the feasibility and front end development phases of the project and that significant funds are only sanctioned when sufficient information has been developed on which to base an informed decision to proceed.

5.6 COST CONTROL TECHNIQUES

A wide range of techniques are used by the cost engineer to identify potential cost problems on a project. These techniques vary depending on the type of project expenditure, and while this publication does not attempt to detail all of these various methods of control it is useful to appreciate some of the more commonly used methods of approach.

5.6.1 Control of design costs

Design costs, encompassing design, procurement and project supervisory activities, can be monitored and forecast by collecting information on progress achieved and man-hours expended and converted to cost by applying recorded man-hour rates.

Information available to the cost engineer on which to base cost predictions will normally comprise the following:

- planned man-hour expenditure by discipline;
- planned progress by discipline;
- actual man-hour expenditure by discipline;
- actual progress by discipline;
- the discipline heads' assessment of man-hours to complete.

Based on this information the cost engineer can derive the man-hours and therefore the cost to complete the design activities, by using the most appropriate of the following three methods:

1. the man-hours required to complete assuming productivity of 100% based on the remaining planned man-hour expenditure;
2. as method 1 but with the result adjusted by the productivity factor achieved to date;
3. using the discipline heads' assessment of man-hours to complete.

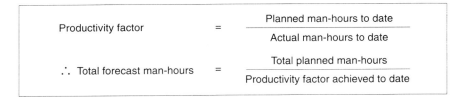

Productivity factor	=	$\dfrac{\text{Planned man-hours to date}}{\text{Actual man-hours to date}}$
∴ Total forecast man-hours	=	$\dfrac{\text{Total planned man-hours}}{\text{Productivity factor achieved to date}}$

Figure 5.3 Use of productivity factors in forecasting

The productivity factor in method 2 is achieved by dividing the man-hours anticipated to be used in achieving an amount of work divided by the actual man-hours expended in achieving the same amount of work as shown in Figure 5.3.

Care should be taken when calculating the productivity factor to ensure that an apparent fall in productivity is not caused by changes not yet raised or agreed which have therefore not been incorporated into the current control estimate. When using this derived productivity factor to forecast man-hours to complete, consideration must be given to factors which may have influenced productivity to date (for example man-hours expended on familiarization) and whether or not they will influence future performance. Based on this assessment it may be necessary to adjust the factor accordingly to obtain a more realistic forecast.

The cost engineer's adjudged forecast of total hours should be based on the principles that during the early stages of a project the forecast will be the man-hour variance to date added to the revised control estimate. Between this stage and latter stages of design the forecast will be the revised control estimate divided by the productivity factor described above. During the latter stages of design the engineer's forecast of hours to complete will be added to the man-hours expended as the forecast of total hours.

By undertaking this assessment for each design engineering discipline on a regular basis it is possible to forecast the final man-hour expenditure and therefore cost of the total design effort for a project as well as to identify problem areas and potential overruns to budget.

5.6.2 Control of equipment costs

It is normal for the control estimate to be based on a reasonably accurate equipment listing produced during the early design phase of the project. It is also normal for the estimator to have obtained budget quotations from suppliers on which to base his assessment of equipment costs included within the estimate.

It is necessary for the cost engineer to check the validity of these estimates as tenders are received for the purchase of these various equipment items. Cost overruns to the estimate provision should be immediately identified and reported, in order that corrective action can be considered and implemented. Reasons for overrun to budget may include but are not limited to the following:

- more exotic materials specified than anticipated in the budget;
- increased capacity or size – occasionally as a consequence of over-design;
- unnecessary requirements written into the specification;
- poor selection of bidders;
- foreign currency fluctuations.

This list gives some indication as to where to start looking for the reasons for the overrun and it is evident that corrective action can be taken in certain cases. It is important to note that when investigating overruns the possibility of a low estimate should not be excluded; but this should be considered only after all avenues of corrective action have been exhausted. It is a psychological fact that once it has been admitted that the budget is low all efforts towards corrective action will cease.

5.6.3 Control of bulk material costs

Assessments will have been made on the quantities of bulk materials required to construct a project. The costs of these bulk materials will have been derived by the estimator using either data from previous projects, with appropriate escalation added, or budget quotes. It is necessary to monitor and refine these assessments throughout the project design phase.

In order to achieve this the cost engineer normally undertakes reviews of quantities of the various materials required to construct the project, based on the drawings and material take-offs produced by the design team, as the design progresses. Close attention has also to be paid to the unit cost of materials as purchasing commences on the project. The findings of such reviews are compared progressively with the relevant allowances in the control estimate. Adverse trends are thereby identified sufficiently early for the project manager to arrange for possible corrective action to be taken.

It is important to note that any adverse trends identified with regard to the quantity of materials required for a project will have a corresponding 'knock on' effect with the cost of installation of those materials.

5.6.4 Control of construction/installation contractor costs

When reviewing the activities of a cost engineer during the construction phase of a project it is necessary to distinguish between the owner's cost engineer and the contractor's cost engineer. The activities on the owner's side will be dictated by the contract strategy adopted on the project, while the activities of the contractor's personnel will depend on the nature of the work for which they are contracted.

The level of control necessary by the owner's cost engineer will largely depend on the basis of the construction contracts let on the project. The basis of these contracts may vary from totally reimbursable to lump sum without variations. Obviously if a contract is let on a lump sum basis without variations there is little requirement for the owner's cost engineer to attempt to analyse performance of the contract, the final cost of that contract being known from the outset. At the other end of the spectrum, with a totally reimbursable contract the owner's cost engineer will be required to put in place the necessary control mechanisms in order that he can predict the likely final cost of the contract.

It is fair to say that whatever strategy is adopted the owner's cost engineer has to take a broader view of the project cost overall, whereas the contractor's cost engineer is only concerned with costs pertinent to that specific contract. One of the responsibilities of the owner's cost engineer is therefore to obtain cost data from the contractor's cost engineer, verify that these are realistic and build them into his overall cost status report for the project. The owner's cost engineer will therefore be required to liaise closely with the contractor's cost engineer obtaining the required information in a format that is compatible with his cost reporting structure and methods.

An important duty of the owner's cost engineer is to ensure that the contractor's cost forecasts and programme predictions are consistent. In all construction activities cost and time are closely linked. In many cases these two functions are handled by separate sections within the contractor's organization and often there is little correlation between the two sets of data. The owner's cost engineer is ideally placed to undertake these fundamental checks.

A further key activity of an owner's cost engineer when assessing construction contractor costs is the recognition of potential claim situations occurring during the construction phase. These assessments are largely based on dialogue with other members of the project team to ascertain what problems are being experienced by a particular contractor on site. If these problems are the owner's responsibility (i.e. late design or late issue of free issue materials), and the contractor's costs are affected as a result, it is likely that a claim for additional costs will be made. The owner's cost engineer must be aware of these potential problems and make adequate allowance for them in the forecast cost to complete.

The contractor's cost engineer will normally have access to more detailed contractor cost and productivity information than the owner's cost engineer and hence will use these data to try and maximize the contractor's performance during the execution of the contract. Many construction organizations are schedule driven and fail to appreciate the productivity costs that go hand in hand with overtime or the cost of equipment retained on site in case it is needed to avert a future schedule delay. The contractor's cost engineer therefore attempts to ensure that labour and construction equipment are used prudently in order to maximize a contractor's profit as well as completing the contract within programme.

The contractor's cost engineer is primarily concerned with the following key elements of the contractor's costs:

1.   direct labour costs;
2.   contractor's material costs;
3.   subcontract costs;

4.   field labour overheads, including:
   * temporary construction costs;
   * consumables;
   * supervision costs;
   * construction tools and equipment.

The contractor's cost engineer employs a number of techniques to control expenditure on the above items. Perhaps the most important technique used by the contractor's cost engineer to derive cost forecasts is the assessment of construction progress on which to base forecasts.

Construction progress is normally summarized and reported as percentage complete. To derive the overall percentage complete it is necessary for the cost engineer to break down the labour elements of construction work into major activities, for example earthworks, concrete work, structural steel, piping installation, etc. The contribution of each of these activities towards the total percentage complete is proportionate to the estimated man-hours necessary to complete that activity as a percentage of the overall estimated man-hours. For example, if the estimated man-hours for placing concrete represented 15% of the overall estimated man-hours and this activity was assessed to be 50% complete the contribution to the overall percentage complete would equal 50% of 15%, i.e. 7.5%.

The assessment of percentage complete by activity is undertaken by the physical checking of construction work accomplished. By comparing the value of work done in estimated man-hour terms with actual man-hours expended on these activities, it is possible to assess productivity achieved to date and therefore forecasts to complete, in much the same way as with design engineering. This technique, widely used in the industry, supplies vital feedback to the project planners for progress updates as well as identifying scope increases and/or productivity problem areas to the project management team.

The paragraphs above are only indicative of the various cost control activities undertaken by a cost engineer on a typical project. These techniques have a variety of names including 'trend analysis', 'sampling', 'unit cost analysis', etc. and vary from one organization to another depending on the sophistication of control systems in place. They all, however, to some extent rely on logical mathematical processes, but above all else communication with and feedback from others within the project team, in order that realistic cost predictions can be made based on all the relevant information available.

5.7 CONTINGENCY MANAGEMENT

The purpose of any estimate is to forecast the final anticipated cost, and since unknown or unspecified costs are always incurred a provision has to be included within an estimate for such unknowns.

A contingency allowance is therefore to cater for unknowns alongside the known scope of the project. It is not an allowance to cover scope changes arising during the course of the project which were not known or envisaged at the project definition stage. The reasons for requiring a contingency allowance and its assessment are covered in Chapter 3 on estimating.

Ownership of contingency is an issue which should be clarified at the commencement of any project. In some cases the owner will prefer to be the holder of this provision and will control its allocation during the execution of the project; in other cases the project team is given the responsibility. Whatever form of approval is adopted it is important that all parties understand how contingency is to be handled and what procedures are required to be followed to ensure that it is managed in a controlled manner.

The contingency allowance can be a large sum, with some demands on its cushion being made early, and others much later. Additional information will become available or modifications envisaged throughout the period of the project, and judgements will have to be made as to whether such items are covered by the contingency allowance or whether the items constitute an addition to the estimated final cost. This judgement can only be made if the contingency sum has been broken down into various defined categories and the calculation of the contingency undertaken in a structured manner by allocating different percentages to each main component according to the quality of the data available at the time the estimate is compiled.

It is only by so doing that control can be maintained, and a judgement made to avoid the running down of the remaining contingency below a level required to accommodate future design development.

By monitoring the various demands on the contingency allowance and using good techniques of cost control a realistic final estimate can be maintained and a contingency allowance retained for the later activities.

5.8 ESCALATION

Due to the excessive time scale of many projects cost escalation is a large element of the budget. Control requires that an understanding exists of the current financial position which provides a platform for the projection of future costs. Differing parts of a project will be designed, purchases made and construction executed at widely varying times, and if escalation is simply a large figure to be used to draw money as required the early problems will be disguised and later items left with an inadequate budget.

The calculation of escalation is dealt with in detail in Chapter 3 on estimating. It is normally held as a separate budget within the estimate, to be allocated in an appropriately controlled manner as the project progresses.

The problem is in obtaining an accurate figure for actual escalation included within a particular cost element of a project and therefore being able to transfer a realistic sum from this separate budget. The cost engineer relies on a number of techniques to make this assessment, one of them being the use of cost indices published by, among others, the Association of Cost Engineers.

It is not intended that this book should address these various techniques; suffice to say it is a budget, as with contingency, that has to be managed and that it will obviously diminish as the project progresses.

5.8.1 Currency exchange rate variations

When involved in construction projects overseas or making purchases of materials, equipment or services from outside of the host country, the project concerned will be exposed to currency exchange rate variations. This exposure, which can be either detrimental or beneficial to the final project cost, is handled differently by companies across the industry.

One method of reporting these variations is to set predetermined exchange rates at the commencement of the project; these are normally derived at the estimating stage. Any variations to these predetermined rates experienced during the execution of the project are then reported separately within the cost report. A benefit of using this method of reporting is that true comparisons of actual cost against the relevant allowance within the control estimate can be made.

As a method of removing the uncertainty of exchange rate variations some owners within the industry choose to buy forward foreign currencies. This involves a contract to buy a specified amount of foreign currency at a specified rate at a specified date in the future. This is a highly specialized area and should only be dealt with by experts in this field.

5.9 THE COST CONTROL REPORT

A cost control report should be issued regularly, normally on a monthly basis during the life of a project. The purpose of the report is to present the up-to-date status of actual costs incurred, commitments entered into and forecasts to complete the project scope.

The report compares the above with the current control estimate (the control estimate plus agreed changes) and highlights differences between the anticipated final cost (AFC) and the estimate provision.

The cost control report together with the updated project programmes are the main means for effective project control in that they gather together information, analyse it and report on the current project status with suggested corrective actions.

The content of the cost report and amount of detail within varies depending on the size of the project and the recipients of the report. Typically a cost report would contain the following, which are further described below:

- summary narrative;
- cost summary tabulation;
- summary of changes to the work;
- various charts and graphs depicting cash flow, cost trends, etc.

5.9.1 Summary narrative

The summary narrative highlights and explains major changes to the cost forecast in the period as well as identifying potential problems and sug-

Project no.
Project title:

COST REPORT SUMMARY

Code	Description	Original control estimate	Approved scope changes	Current control estimate	Value of work done	Commit-ment to date	Estimate to complete	Anticipated final cost	Over/ (under) budget
1100	Major equipment	184.2	7.0	191.2	170.3	180.3	15.0	185.3	(5.9)
1200	Piping	291.0	3.6	294.6	275.6	302.8	44.6	320.2	25.6
1300	Civil work, building and steelwork	324.0	13.8	337.8	373.7	372.7	0.0	373.7	35.9
1400	Electrical	135.0	1.3	136.3	35.6	54.5	100.7	136.3	0.0
1500	Instrumentation	210.0	6.7	216.7	100.2	106.2	116.5	216.7	0.0
1600	Insulation	40.0	0.0	40.0	10.0	25.0	15.0	25.0	(15.0)
1700	Painting	15.0	0.0	15.0	3.5	12.5	9.0	12.5	(2.5)
1800	Temporary construction facilities	25.0	0.0	25.0	15.8	22.4	9.2	25.0	0.0
1900	Spares	5.0	1.0	6.0	0.0	0.0	6.0	6.0	0.0
2000	Fire protection	100.0	0.0	100.0	36.3	76.5	43.7	80.0	(20.0)
3100	Engineering	220.0	14.6	234.6	214.2	214.2	39.8	254.0	19.4
3300	Professional fees	58.0	0.0	58.0	33.0	34.3	11.3	44.3	(13.7)
	Subtotal	**1607.2**	**48.0**	**1655.2**	**1268.2**	**1401.4**	**410.8**	**1679.0**	**23.8**
5000	Contingency	181.0	1.5	182.5	0.0	0.0	158.7	158.7	(23.8)
	Total project costs	**1788.2**	**49.5**	**1837.7**	**1268.2**	**1401.4**	**569.5**	**1837.7**	**0.0**

Figure 5.4 Example cost report summary

gested corrective actions necessary to avoid them. It is normal that separate narratives are included for the main elements of the project, which are:

- design;
- materials and equipment;
- fabrication/construction;
- (other) subcontracts.

5.9.2 Cost summary tabulation

The key document in any report is the cost summary tabulation. The format and level of detail of such a summary varies across the industry and is often adapted to meet the circumstances of the particular project. An example of how this may be presented is shown in Figure 5.4.

The cost summary tabulation typically displays the following information at levels of detail to suit the users of the report and may be supplemented by a scope change summary, various reports, charts and graphs:

- original control estimate;
- approved scope changes;
- current control estimate;
- value of work done;
- commitment (to reporting date);
- estimated costs to complete;
- anticipated final cost;
- variances between the current control estimate and the anticipated final cost;
- expenditure to date.

By using the applicable coding structure, the cost summary tabulation can be presented in accordance with the cost breakdown or work breakdown.

5.9.3 Scope changes summary

The scope changes summary is a list of both scope changes approved together with their impact on the original control estimate and pending scope changes currently under consideration.

5.9.4 Charts and graphs

Charts and graphs may include the following:

- design progress reports and trend analysis;
- progress curves;
- productivity profiles;
- manpower histograms;
- short- and long-term cash flow (expenditure) forecasts (see Figure 5.5).

Computers allow for a vast quantity of detailed information to be stored and manipulated, making it a simple task to produce a large imposing-looking report containing information on both work packages and cost codes to a very detailed level, but hiding problem areas in a jumble of figures that nobody in authority has the time to read. It is recommended that the cost control report is kept as brief as possible, with reports being summarized to suit the needs of the recipient. Higher level reports should be prepared for those who must make the higher level decisions, while more detailed information should be made available to those who need to deal with the detail.

5.10 OTHER REPORTS

Other reports which the cost engineer may be required to issue at regular intervals comprise the following.

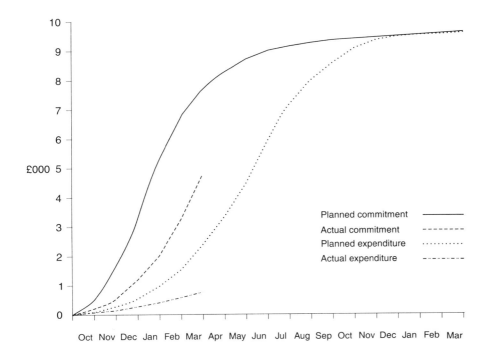

Figure 5.5 Expenditure and commitment curves

5.10.1 Exception reports

As a means of rapidly identifying problem areas where progress is not satisfactory, or man-hour expenditure excessive, exception reports will be produced on a frequent regular basis throughout the life of the project.

The exception reports will list those activities, or individual documents, where predicted completion dates are later than the activity early finish date or where an excessive man-hour/cost variance exists.

5.10.2 Cash flow forecasts

A cash flow forecast will frequently be requested by the project which will be produced by effectively pricing the project schedule. This is achieved by spreading the expenditure against the periods over which that expenditure is being incurred in line with an appropriate expenditure profile, in order to produce an anticipated cash flow monthly expenditure for the remainder of the project.

There is often a requirement for both long- and short-term forecasts, usually covering the subsequent three months. The short-term forecast would be used by the financing group to draw money down from the owner or co-venturers in readiness for the payment of monthly invoices.

Part 2 Planning and progress control

5.11 INTRODUCTION

One of the fundamental elements of project control is the planning activity. Control requires an understanding of the current status, the ultimate goal and of the actions necessary to achieve that goal. This is as true in the context of the overall project as it is in the day-to-day activities of each person working on a project.

Planning itself comprises all activities involved in setting up the strategy for a project; it influences the method of execution, the resourcing, the procurement policy and the contracting strategy. The planning function continues throughout a project, monitoring progress, forecasting future trends, measuring results, and reporting status.

5.12 THE IMPORTANCE OF PLANNING

The importance of project planning can best be illustrated through previous project experience. Many projects fail due to misconceptions, underestimations of time and the like in the initial planning stages and due to the failure of management to ensure that the plan is adhered to. It should be understood that planning engineers plan the project and project managers manage the project. It is therefore the duty of the planning engineer to establish the project activities and events, their logical relationships and interrelationships and the sequence in which they are to be accomplished, which is the route map by which the project manager will direct the project. The failure by the planning engineer to produce the correct chart or the project manager to follow its requirements will have dire consequences.

The rectification of problems may be comparatively painless if identified at an early stage; but as a project progresses the cost of rectifying error or incorporating change increases dramatically. The completion date is also increasingly at risk and the consequential lack of time in the later stages of a project can result in restrictions, or modifications to the preferred strategies, uneconomical working and additional cost.

If work cannot be anticipated it cannot be planned, if work cannot be planned it cannot be controlled and will be undertaken in an *ad hoc* manner with costs being monitored and not controlled. This comment applies equally when the word 'time' is substituted for 'cost'.

The provision of a properly considered plan is vital to the commercial success of a project and must be produced by those who have experience of a similar type of project, and who appreciate the full range of dependencies and relationships between the various activities which have to be undertaken in order to complete the project.

5.13 ORGANIZATION

To undertake a complex project successfully, it will be necessary to establish a dedicated, responsible, planning team providing a comprehensive planning, monitoring and control service that gathers and distributes information by the receipt and issue of appropriate reports through the whole project, from the owner to the project team and through the design and construction contractors to the various subcontractors and suppliers. The degree of transparency and control that can be exercised at any level will depend on the contractual relationships. It will be of paramount importance to the project that the planning engineer ensures that a condition of 'no surprise' exists throughout the life of the project.

While the planning activity embraces the time span from the moment a project is born through to its completion, the start point of a particular planning exercise will depend on whether the planner is working for an owner, contractor or subcontractor. A planner working on behalf of an owner will start with a blank sheet of paper and will be restrained only by the degree of urgency expressed by his management. Planners working for contractors or subcontractors, on the other hand, will work within increasing constraints the lower they are on the contracting pyramid.

The control of a complex project may require the owner's project team to insist that all participating contractors use planning systems that are compatible with the higher level requirement, both in terms of technique and system software. The same team may require information to be provided on progress achieved, anticipated completion dates for the work or parts thereof, and perhaps manning levels to ensure the smooth progression of work from one contractor to another. In addition the team may require planning and progress systems to be compatible with work packages and other work breakdown structures to provide integration of reporting. Such requirements can, of course, have a profound effect on contract strategies and relationships.

5.14 LEVELS OF PLANNING

The degree of detail required by the various personnel or organizations working on a project depends on their position within the project. What may be considered as mere detail to the owner may be the full scope of work to a subcontractor.

As with other areas of project services, planning systems and reporting requirements are arranged in a hierarchy, each level of the hierarchy containing detail appropriate to the work being controlled and level of management involved. The level of detail necessary to control the day-to-day activities is not required by a project manager reviewing the general status of the project, while corporate management may only be interested in a high-level 'executive summary'.

Various levels of planning and reporting are therefore necessary to reflect the requirements of the users of the information, with each level being capable of interrogation at a greater level of detail through the hierarchical structure.

The planning function commences by the setting up of a plan for the overall project in the form of a Gantt or bar chart. The project programme is then prepared which sets the plan in a time-scale and takes account of key activity dates and milestones. Detailed control programmes are then produced in bar chart format derived from logic networks. Logic networks identify activity durations, the logical links between activities and take account of the salient aspects of the required and available resources. Finally, individual task level schedules are prepared, based on the control programme, for the day-to-day control of the physical work.

Ownership of the plan is not confined to the planning engineer and it cannot be emphasized too strongly that the advice and participation of the parties that will be working the plan, e.g. engineers, buyers, construction managers, etc. should be solicited. All members of the project team should participate in the formulation of the plan from its inception and fully 'buy in' to the schedule in its final form.

Physical work will be expressed in terms of physical progress achieved. Progess related to design and similar activities will frequently be expressed in terms of deliverables. A deliverable is the physical result of the completion of a discrete package of work and usually relates, but is not limited, to drawings, specifications, procedures, reports and other documents.

Described below are the four typical levels of planning, there may be additional levels depending on the nature of the project. Levels 1 and 2 will be produced by the owner's project team and levels 3 and 4 by the contractor, who may also have his own version of levels 1 and 2.

5.14.1 Level 1 bar chart

The level 1 bar chart is the most basic graphical demonstration of work plotted against time. It will show the overall programme for completion of the project in the form of a time based bar chart (sometimes called a Gantt chart) of perhaps 30 activities and in single sheet format.

Such a chart, a simplified version of which is shown in Figure 5.6, provides a very clear overall view of a project's requirements with completed work perhaps represented as a solid line and outstanding work as a broken line compared to a 'time now' line to show activities ahead or behind programme. A bar chart cannot indicate the complex interrelationships and dependencies between the many activities in a project and is not a document against which detailed control can be exercised since it does not contain the information necessary to monitor progress in controllable portions.

The level 1 programme will typically be used as a summary by senior management and may also contain summary progress information and key dates as shown in Figure 5.6.

Key dates, sometimes referred to as milestones, will typically have a single deliverable for their achievement, for example:

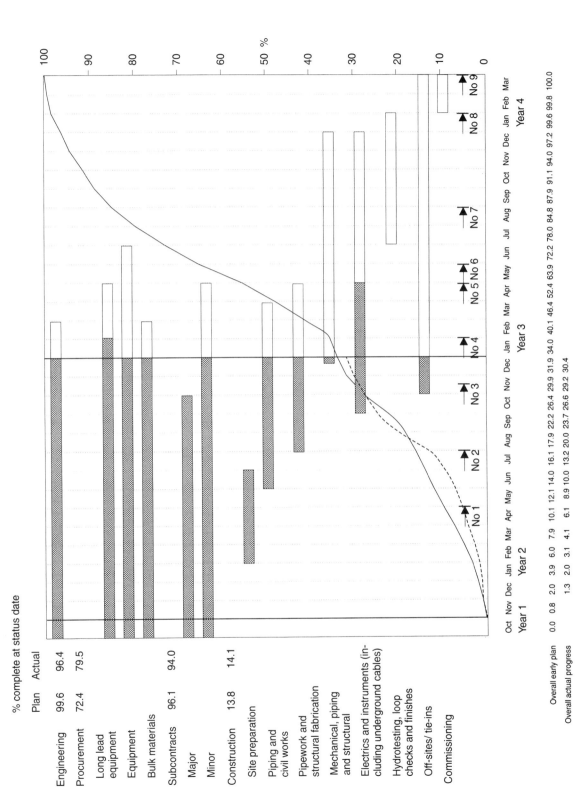

Figure 5.6 Level 1 bar chart with key dates (milestones) and progress curve

- placement of purchase orders for particular items of equipment;
- mobilization of a key contractor;
- completion of detail design;
- delivery to site of certain specified items of equipment;
- completion of heavy lifts;
- power on;
- mechanical completion.

The key dates will result from the more detailed considerations in the level 2 programme.

5.14.2 Level 2 programme

The level 2 programme is the highest working level of the planning structure it is the level at which the relationships between the various activities are identified and will be the project manager's working tool. Since it will identify all key activities and key dates it will typically be the project manager's 'master schedule' and can be used to track predetermined project milestones and the related approvals required from the owner.

This programme will be in the form of a time based linked bar chart of perhaps 100 key activities and stages of completion similar to that shown in Figure 5.7. All dates shown on this programme having been derived from the analysed level 3 network.

The level 2 programme will show the overall project status and will be updated and reissued at regular intervals showing, relative to a vertical dateline, the progress planned and achieved for each activity.

5.14.3 Level 3 network

The level 3 programme is usually viewed in the form of a bar chart derived from a logically linked network.

A network identifies all activities necessary to the completion of a project, the time required for each activity and the logical dependency of activities on each other. For each network there will be a chain of logically linked activities from the start of the network through to the finish of the project that defines a duration through the project as a whole which cannot be reduced. This is called the critical path because any overrun on any of the activities on this critical path will cause an overrun of the project (see critical path and float section 5.16).

A first pass at the level 3 programme, in the form of a trial network, will probably be undertaken by the owner's project team, but the 'definitive' level 3 programme is usually produced by the contractor as his detailed control programme. This programme covers the entire scope of the contractor's work in the form of a logically linked network, identifying the key dates, deliverables, interfaces, material procurement and (sub-) contracting activities.

In producing the 'definitive' level 3 programme, the contractor's planning engineer will take full account of all 'softer' links such as resource

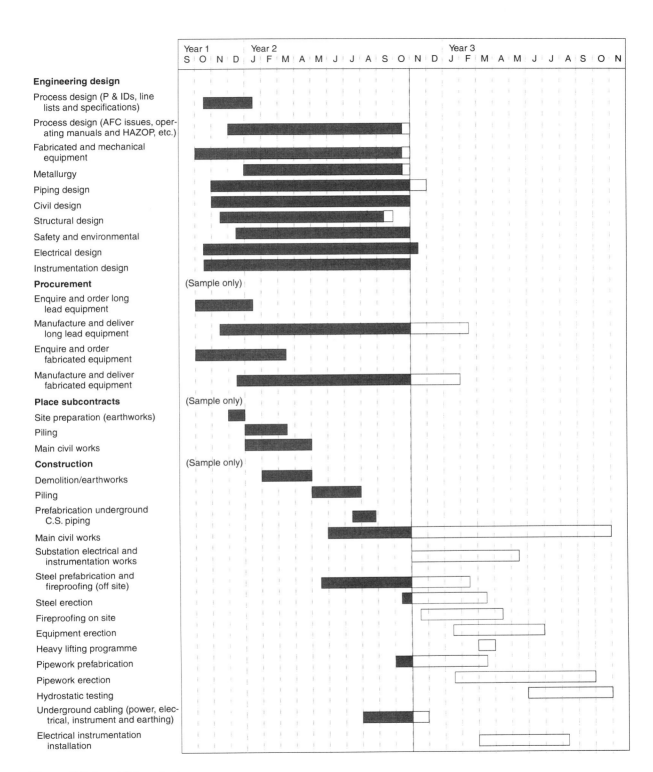

Figure 5.7 Level 2 bar chart programme

and manpower constraints, which are under management control, as compared to logical links which are not. Thus, for example, the stripping of shutters is absolutely dependent on the curing of concrete, but it may also be dependent on the availability of carpenters. The first link cannot be altered, whereas the second can.

The programme will allow for the requirements of any significant key dates in the 'project calendar' and when the analysis is performed (see section 5.16), it calculates the start and completion dates, taking account of non-working days and 'total float' for each activity. Float is the amount of overrun on an activity that can be accommodated without it affecting the project completion date. Activity durations are usually shown in days.

When the detailed plan has been developed and the analysis run and agreed, activities are loaded with man-hours and preliminary manpower histograms produced. These are reviewed with the discipline lead engineers, construction managers, etc. and activities are moved within their float limitations to obtain the optimum 'smoothed' histogram, i.e. a labour profile without excessive peaks and troughs. At this point the programme is frozen, often referred to as being 'base lined' and the level 2 programme and level 1 bar chart are then produced.

The level 3 programme is the basis against which those undertaking the work report progress and likely completion dates for each activity, using summated information obtained from level 4. The level 3 programme will usually be updated fortnightly during design increasing to weekly during the construction phase. The update will be based on information received from engineers, construction managers, etc. indicating achieved progress and current forecast against the plan for each activity.

The contractor should be allowed the freedom to plan and manage his own work and unless the owner is carrying out some of the work himself, level 3 will be the lowest level at which the owner's project team will operate on a day-to-day basis. The owner will require such detail as will allow him to manage the various interfaces, but should resist the temptation to plan at too low a level and in too much detail.

5.14.4 Level 4 schedules

The level 4 schedules convert the theory of the network into the practical considerations for executing the work. Level 4 information generally comprises registers containing all the relevant details of deliverables or site work pertaining to the work elements within each level 3 activity and are the basis for the subsequent day-to-day planning, monitoring and control of workscope and physical progress. The level 4 schedules are produced by or on behalf of the contractor or whoever is chosen to execute the work.

Every level 4 design deliverable or element of site work, identified by the design disciplines or construction managers, will be assigned a code number which will link the documents or work to activities at level 3.

It is usual for the activities at level 4 to be coded and organized so as to be capable of being summated in a number of different ways. For example,

JOBCARD NO	DISCIPLINE		PAGE NO	
CONTRACT NO	ISSUE DATE			
REVISION NO	REV DATE			
PLANNED START	PLANNED FINISH		DURATION	HOURS
COST CODE				
JOB DESCRIPTION e.g. fabricate and weld 'can A' jacket leg A1	RESOURCE TYPE	RESOURCE QUANTITY	RESOURCE DURATION	

DRAWINGS REQUIRED

SERVICES REQUIRED · WELDING DOC REF NOS

e.g. SCAFFOLDING, RIGGING · TOOLS, NDT

PERMITS REQUIRED
HOT WORK · VESSEL ENTRY · ELECTRICAL

CONTRACTORS' SIGNATURES

INSPECTOR / DATE · SUPERVISOR / DATE

SITE MANAGEMENT SIGNATURES

INSPECTOR / DATE · SUPERVISOR / DATE

Figure 5.8 Job card example

although initially the reporting and planning of level 4 activities may be on an area basis, at a later stage it may be necessary to produce information sorted on a system basis so as to provide a detailed report on the activities and man-hours required to complete each of the process systems regardless of area.

5.14.5 Shutdown planning

When a process plant is closed down for repair, maintenance or upgrading (see Chapter 11 on operational maintenance), construction activities are planned and controlled vigorously due to the extent of the work to be undertaken over a very short contract period, which frequently requires 24-hour shift working.

In these situations job cards similar to that shown in Figure 5.8 may be used. The job card will contain, for example, the degree of detail necessary for a field supervisor to organize his workforce, since it will designate the timing and resources allocated for the completion of each activity. It will also carry details of all drawings applicable to the work.

5.15 NETWORK TECHNIQUES

The network is at the heart of the planning function in that it contains the logic on which the various planning schedules are based.

A network consists of a sequence of events logically linked together and contains information regarding the nature of events, the order in which events can be achieved and the duration of any activity necessary to their achievement. The duration of a path through a network is therefore dependent on the duration of the constituent activities, whereas progress is plotted against events.

There are two conventions of representing networks. The first advocates activities as arrows, the event being the node at which the arrows meet. The second is termed a precedence network which represents the activities as the nodes of the diagram which are connected into a chain of logic by arrows called constraints The former is sometimes referred to as 'activity on arrow', the second 'activity on node' which is current practice.

5.16 CRITICAL PATH AND FLOAT

In order to calculate the earliest time of an event, it is necessary to consider the durations of activities that precede it. This is called a forward pass.

In order to calculate the latest time that an event can occur while not affecting the overall completion date, it is necessary to consider the durations of activities between the event and the end of the project. This is called a backward pass.

The forward and backward passes through the network produce for each event a range of times at which the event can take place without jeopardizing the project completion. The limits of this range are the earliest possible event time, produced by the forward pass and latest possible event time, produced by the backward pass. Where the earliest and latest dates coincide these activities are known as critical activities and the critical path is that which links these critical activities.

When the forward and backward passes have been completed an end date for the project becomes evident. By fixing this end date any future analysis extending critical activities will be indicated as negative float, which highlights the offending activities to the planning engineer. To retain the original contract completion dates it will be necessary to review the (now) supercritical activities and perhaps increase manning levels to reduce the duration of the activity, or manage the problem by highlighting the difficulty and giving it regular specific attention.

Activities may be on the critical path at a particular time in the project, but may thereafter cease to be critical, due to a particular happening, or lack of happening; similarly, activities that were not on the critical path may become critical.

Events away from the critical path have a range of available times calculated by the forward and backward pass, the difference between the earliest and latest times for an event is known as float. The critical path will be the path with the least float, normally zero.

For a definition of the terms used in project networks reference should be made to British Standard (BS 4335) entitled *Glossary of Terms used in Project Network Techniques*.

5.17 PROGRESS MEASUREMENT

In order for the project team to forecast the final completion date and cost, a basis for comparison needs to be established between the anticipated and actual progress and anticipated cost and cost to date.

The level 4 schedule forms the basis of the contractor's progress control system. It is at this level that the contractor will record progress, identifying what has been achieved and when it was achieved. The actual hours used in its achievement will be recorded against time sheets which are normally the equivalent of level 3 information.

This information will be used to forecast man-hours required to complete the work taking into account progress and productivity achieved to date, by area, discipline or system.

Progress is not measured by comparing time expended against total time allowed, nor is it measured by comparing man-hours expended and total man-hours anticipated, or by comparing amount paid and total value of the contract. Progress is a comparison between work anticipated and a measurement of actual physical work achieved. Progress will only be measured against 'direct' or 'progressable' activities, i.e. activities that directly contribute to the production of the unit. Management and support trades, for example, are not measured as progress.

Measurement is carried out periodically at agreed cut-off dates, the method of physical measurement of progress varying between disciplines. Clearly the method of measurement of design, procurement and similar activities will differ from that used in connection with construction works where physical progress is directly measurable.

In order that progress on design and similar activities does not remain a matter of opinion it is necessary to agree, prior to commencement, the stages through which deliverables, e.g. layout drawings, general arrangement drawings, enquiry documents, etc. will proceed in order to establish a framework against which progress can be evaluated. This is usually referred to as benchmarking. The agreed stages should contain a description of the 'quality' of the deliverable required, as progress on a deliverable containing various 'holds' or blanks is questionable.

Original design effort and progress are based on the designer's estimate as to the types and quantities of documents to be produced and man-hours established for the different stages through which the document has to be processed. Progress is assessed by the engineers at the lowest level of detail, e.g. drawing, specification, study, etc. The physical status of the document is reviewed using the benchmarks laid down for the stages, per document and the engineer's own judgement.

Procurement can be planned and monitored by setting a procurement cycle which includes such activities as enquiry issue, order placement, etc. and allocating man-hours accordingly. Time allocated to each stage of the cycle takes account of vendor (supplier) documentation requirements, preparation of design deliverables, etc. Procurement progress is measured against the procurement cycle per requisition for the total procurement effort.

Design programmes will usually be construction driven, in that the needs of the large construction workforce will take priority over the inconvenience of the designers. The delivery dates for procured items will be linked to the construction requirements; equipment and materials will therefore be 'reverse engineered' to take account of when it is required to suit site activities. Certain equipment will be critical with regard to vendor (supplier) data rather than delivery, in which case the enquiries will be programmed early in the project so as to be in receipt of data to suit design requirements.

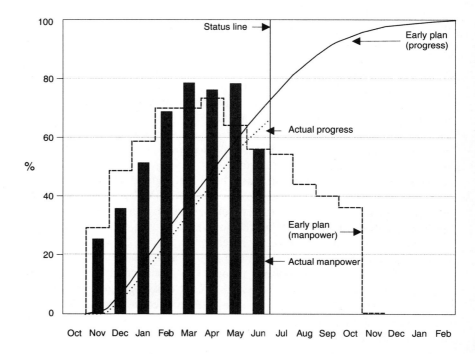

Figure 5.9 Typical level 2 progress curve

Information obtained from the physical progress measurement will be used in conjunction with the 'S' curves and planning breakdown to show actual physical progress against planned progress on a given date (see Figure 5.9).

Measurement undertaken for the purposes of reporting progress is not necessarily suitable for other purposes, i.e. agreeing accounts, although it is frequently used as a basis for the agreement of interim payments by the reported progress being applied to the current value of an element.

5.17.1 Histograms and progress 'S' curves

Histograms indicate the manner in which labour levels periodically increase and decrease over the life of a project or part thereof. They fall into two categories, progressable and non-progressable as follows:

1. non-progressable:
    *   management and support services;
2. progressable:
    *   engineering design;
    *   procurement;
    *   supervision;
    *   staff;
    *   labour by trade.

There may also be a summary histogram indicating overall labour.

The contractor is usually required to submit manpower histograms for the work together with curves ('S' curves) showing graphically the anticipated progress versus actual progress throughout the contract period (see Figure 5.9).

The 'S' curves will show the planned and actual percentage completion for each reporting period and will be generated from the level 4 programme. The initial planned curves usually remain unchanged for the duration of the contract unless there is a major change that would otherwise make comparisons irrelevant.

Histograms can be determined by applying resources to each network activity and aggregating the results per period in order to obtain required labour totals weekly, monthly, or to whatever other period may be chosen. Computerized network programmes not only carry out the aggregation but also plot the resulting histograms. Histograms may disclose erratic labour levels that may indicate that the network logic requires attention.

5.18 CONTINUING CONTROL

Once the project is under way the planning engineer will regularly review and report on the status of the work and will update the planning database. The planning engineer will coordinate his activities with the cost and contract engineers to ensure that all project controls can be used in conjunction each with the other. This is particularly important with regard to the incorporation of agreed changes into the scope of work.

Information will be collected on man-hour expenditure and progress achieved for comparison with that planned. The planning engineer will, together with the material controller, monitor deliveries of material and equipment against planned and actual requirements together with the issue of design information and the release of any 'holds' on design drawings.

5.18.1 Progress reports

Each contractor actively involved in the project will produce progress reports at regular stipulated intervals, in accordance with the requirements of the project team. The project team's planning engineer will then produce a consolidated progress report for the whole project.

A typical monthly progress report required of a contractor will comprise the following.

Summary

A summary of the current status of engineering design, procurement, contract placement, equipment and material deliveries and construction progress. Specific reference should be made to any current problems, bottlenecks, delivery lapses, etc., the effect on programme and/or costs, and an indication of the steps being taken to overcome them.

Achievement

A description of progress and summary of the labour force by trade or discipline for both the contractor and his subcontractors.

Objectives

This comprises a description of activities to be carried out in the next period.

Areas of concern

Highlights any problem areas that may cause delay to a key date or to the completion of the work.

Pending or approved change orders

A summary of pending or approved change orders obtained from the cost engineer or contract engineer.

Accidents and losses

A summary of accidents and losses as produced by the safety officer.

Planning matters

A summary of progress indicating any departures from the agreed programme and their likely effect on overall completion of the work.

Bar charts or other graphical presentation to show planned and actual progress for:

- engineering design;
- procurement;
- material deliveries to site;
- construction.

The graphical presentation is frequently in the form of a histogram based on original control man-hours and the original project programme and shows current manpower requirements, overall and by discipline. The basis of the histogram will be updated periodically based on revised man-hours and agreed revised programme.

The cost engineer will use information relating to progress achievement when forecasting final cost and developing other statistics; it is important that cost and planning controls are maintained on a directly comparable basis.

5.18.2 Intermediate reports

On major projects or during critical stages of a project it may be necessary for a contractor to be required to produce weekly or even daily reports. Such reports may be restricted to the following subjects:

- details of the actual compared with the planned physical progress of the work achieved during that period;
- manning levels by major activities;
- materials and equipment status, indicating any areas of special concern;
- any matters which could jeopardize timely completion of the work;
- activities to be carried out during the following period;
- weather report;
- all downtime;
- accidents.

Part 3 Document control

5.19 INTRODUCTION

The number of drawings, specifications, data sheets and certificates generated on a major project by the owner, designer, construction contractors, subcontractors, suppliers and the like can run into many thousands, each of which may go through a series of revisions.

Engineering projects are complex and involve large design and construction workforces working in parallel or consecutively on many different disciplines. Each discipline depends on information provided by others and each in turn provides information to others. Such information may relate to space requirements or to the functional or structural dependence of one system on another.

This high level of interdependence, together with the usual tight schedules and large workforces means that failure to produce a document on time, or failure to comment on a document produced by others in accordance with the agreed time scales can cause serious delay and/or costly change to the design and the construction work.

The progress of documents, whether produced by the owner, designer, contractor, or supplier and at whatever location, needs to be tracked through the various stages of design review, comment and issue.

5.20 MAIN ACTIVITIES

Efficient coordination between all the design disciplines and suppliers is vital. Strict control needs to be exercised over the progress of the key deliverables and their availability, in order to ensure that all persons requiring the use of the information have access to the latest revisions at all times.

On major projects it is often necessary to establish a document control team to provide the project with a document management service covering all document control requirements and related functions from contract award to project handover, possibly extending into the provision of manuals necessary to the subsequent operation and maintenance of the plant.

The main activities of the document control team will be to provide an information service to discipline engineers, construction contractors and suppliers, in various locations covering registration, monitoring, printing, distribution, expediting and status reporting on all documents produced by the project.

Although exceptions may be made in the case of documentation of a confidential nature, document control will be the focal point for the registration, monitoring and distribution of project documents.

The team will therefore:

- establish a system of document referencing;
- establish a document tracking system to control and report on the generation, issue, receipt, review and return by each discipline;
- provide a system of document storage and retrieval.

5.21 CONTROL METHODS

Document control will operate at the level at which individual documents can be identified and tracked, which is usually the equivalent of level 4 of the planning system. The document control team will be responsible for establishing and maintaining the document database at this level.

Information flows into the project from subcontractors and suppliers, etc. as well as flowing out from the design team. In order to ensure that the required information is received, in the required format, in the required number and at the required time a document schedule will be issued with each purchase order or subcontract, under which there is a design responsibility, as-built drawing responsibility or the like. Document control will liaise with the originator of the information with regard to format while the expediters will ensure receipt of the required documents at the appropriate time.

5.21.1 Registration of documents

All documents generated within or for the project, will be identified by a unique number to facilitate registration, reference and retrieval.

Registration of each deliverable will record as a minimum:

- document number;
- subject/title;
- date of original and current issue;
- revision/status/date;
- purpose of issue (e.g. for review; for design; for enquiry; for construction);
- file reference.

5.21.2 Transmission of documents

Documents will usually be issued under cover of a hard copy letter, electronic message or other pro forma means of recording its issue.

It is important to consider the wording of the transmittal which should clearly state the date of issue and the reason for which it is issued. If, as in the case of drawings, specifications, etc. being issued to a design or construction contractor, it should be made clear whether the issue is or is not an instruction to proceed with the work shown thereon.

It is also important that the document controller knows not only that drawings were issued but also that they were received through some means of formal acknowledgement. To rely on the contractor spotting a gap in a sequence of transmittal slip numbers is not a viable option when time is of the essence.

5.21.3 Review of design deliverables

Deliverables received from subcontractors and suppliers requiring review by others will normally be distributed to the reviewing departments under cover of an internal transmittal that records issue date and the date by which comments are required. Approval cycles will be tracked and recorded and the document control team will expedite comments not received by the required date in order to maintain the programme.

5.21.4 Distribution

As previously stated, it is important that members of the project work from the latest information. This is especially true in the case of the design team.

Clearly not everybody on a project requires copies of each of the thousands of documents that a major project will produce, and therefore a document distribution matrix will be established at an early stage and continually maintained to meet the requirements of the project, thereby minimizing unnecessary distribution.

The matrix will identify all the documents which it is anticipated will be produced on that specific stage of the work. The document control team will invite each member of the project to identify on the matrix which of the documents they wish to receive as part of the standard distribution. In addition an updated document list will be available at all times, either in hard copy form or on a computer database, against which an engineer can check on the availability of documents that he had not originally identified as being required to receive on the standard distribution list and also to check that he is now using the latest information.

The document control team can ensure that engineers always have access to the most up-to-date information but it is up to the individual engineer to ensure that he is using that which is available.

5.22 AS-BUILT DRAWINGS

The activities necessary for the preparation of as-built drawings may vary dependent upon whether or not computer aided design (CAD) is being used. The contractor will be required to mark up copies of the drawings issued to him for construction purposes, to show any changes made at site in order to ensure that on completion the owner has a complete and accurate record of the work. If CAD is used the information contained on the marked up drawings will usually be transferred into the computer database.

5.23 REPORTING

The status of design deliverables must be known if project progress is to be monitored. Status reviews and reports will therefore usually be generated at the equivalent of planning level 4 using a database.

Prior to the commencement of engineering design, each discipline will review the list of drawings and other deliverables required to satisfy the complete project scope. This listing will be used by the document control team to establish the initial database against which control will be exercised. Planned and forecast dates will be set by the planning engineers and as each deliverable is received by document control, actual dates will be held within a 'live' operating environment, enabling the current status of deliverables to be known and reported. The database will be updated as further information becomes available.

The document control team will need to maintain complete records of documents linked to a reporting system, which will typically allow for:

- tracking of all types of deliverable documents with full revision history – designer and supplier drawings and specifications, requisitions, purchase orders, etc.;
- tracing successions and cancellations;
- logging equipment to documents;
- carrying multiple document numbers for owner, contractor and supplier with variable numbering structures to suit all needs.

5.24 RETENTION OF DOCUMENTS

The amount of paperwork generated on a major project is immense. Some of it is destroyed but many documents need to be retained over significant periods of time.

Different clients and different countries will each have particular requirements regarding ongoing liabilities which must be established prior to the commencement of the project in order that the relevant documents are retained for the requisite period.

Part 4 Material control

5.25 INTRODUCTION

The cost of equipment or materials on an engineering project is one of the most significant cost components in the estimate and the consequential cost of materials failing to arrive at the required time or at the wrong location causes delay and disruption to the construction activities often out of all proportion to the value of the missing item.

Whether the material controller is part of the project management team and dealing with 'free issue' material only, or whether part of a contractor's team supplying all material, the overall requirements of material control remain the same.

The material controller is responsible for ensuring that materials are received:

- at the required time;
- at the required location;
- in the required quantities;
- to the required specification.

The material controller will develop and operate a system that monitors status against these requirements, inspects and approves storage conditions and procedures and records any maintenance undertaken during the construction period, in order to ensure that material and equipment is in good condition when installed and that any guarantee is not invalidated.

5.26 TRACEABILITY

Due to the hazardous nature of process plants, in addition to the commercial considerations stated above, there will frequently be external requirements to satisfy the certifying authorities, insurance companies and statutory authorities with regard to material quality and traceability. Such requirements are in order to ensure that the quality and integrity of the completed plant are to the appropriate standard, that the plant can be operated safely, with a minimal risk of failure and that identification and traceability of supply is facilitated in the event of equipment or material failure.

Traceability is usually required for primary items such as equipment, piping, fittings, structural steelwork, etc., and records are therefore normally maintained through a coding structure that identifies the supplier, batch number and location within the finished project so that if necessary the location of all other similar components from the same batch can be established and the items inspected or replaced.

The means of providing traceability will depend on the nature of the material. For example, electrical cables should have all the required information printed on the cable at regular intervals, while each piece of pipework or structural steelwork must have a heat and cast number stamped on it. Many owners and design engineers now require that such numbers are also on the delivery notes. When a piece of pipe or plate is cut from the original marked item of material, the heat and cast numbers must be transferred on to the piece that is left unnumbered, otherwise that material will not be allowed to be incorporated into the finished unit and will be worth scrap value only.

5.27 COORDINATION

During the procurement phase the material controller will act as coordinator between the construction contractor and the design office, by informing the design engineers of the coding systems to be used for material control purposes and providing expediters with the required on-site dates. The material controller will also seek the efficient use of material by agreeing allowances for contingency and waste with all parties.

Coordination often includes 'nesting'. Nesting is a term used for arranging the various shapes and sizes of steel plate required by the design in order to reduce wastage when they are cut from the standard rectangular plate delivered from the steel supplier. Nesting will usually be undertaken by the fabrication contractor and verified by the design engineers. Once 'nested' the material controller should record nesting through material allocation sheets.

5.28 SITE STORAGE

Materials and equipment that have been correctly delivered, will be moved from the receiving area into site storage. Such storage varies from a lay-down area typically for pipe, steel, etc., to a warehouse for electrical or instrumentation equipment. Note should be taken of any special requirements for security of the materials, especially on a multi-contractor or multi-project site, and for the preservation and protection of the materials such as heated storage. The issue of materials for fabrication and installation will vary between projects but it is usual to requisition materials from the stores as they are required to be incorporated into the work.

Once in the storage area the materials will be under the immediate control of a stores organization who either report to the material controller or work to procedures agreed with the material controller. The stores organization will be equipped with the necessary lifting and transport plant and other items necessary for the safe handling and protection of the materials and equipment.

The material controller is responsible for ensuring that the equipment and material are kept in good condition. This will involve regular inspection of stores and ensuring that equipment is subject to a system of preventive maintenance that may involve regular inspection, lubrication or turning of motors, pumps, etc.

Accurate material control will be achieved by the keeping of records of receipts, stocks and issues and the cross-referencing of quantities on order and delivery forecasts. In this way, material shortage reports can be generated so that prompt action can be taken to obtain additional material, i.e. by expediting effort or the placement of new orders.

Particularly heavy items of equipment will be required to be delivered on a predetermined date and off-loaded directly on to prepared foundations to avoid double handling.

5.29 SPARES

Material controllers will often coordinate the interface between the construction contractors and the plant operations by recording and handing over surplus materials and spares to the plant operations together with all associated certification, etc.

Breakdowns on a manufacturing plant are an insurable risk and the longer a plant is idle the higher the cost to the insurance company.

Consequently insurance companies frequently require that certain key spares are kept in store in order to avoid delay caused by having to order, manufacture and deliver replacement parts. Such spares are often referred to as insurance spares and a material controller must ensure that such spares are handed over to the operations group in good condition and with the necessary documentation.

5.30 CLOSE OUT

A dossier of test certificates will be produced by the material controller for handing over to the owner together with a set of as-built drawings and iso-metrics which contain reference to the heat and cast numbers, etc.

On project completion, the material control records will be used to carry out a material reconciliation in order that all materials are accounted for and costed. Surplus material may then be:

* sold back to the suppliers (via 'buy back' clauses in original orders);
* auctioned to interested companies;
* sold for scrap value.

Part 5 Coordination of procedures

5.31 INTRODUCTION

The professional institutions associated with the UK building and civil engineering industries have developed standard documents and training requirements which allow any person working on any project to have a rea-sonable idea of what his opposite member on another project will be doing.

By comparison the engineering industries have developed from a number of different industries and geographical areas. Although the funda-mentals of administration may be similar in the various divisions of the industry the manner in which it is practised and by whom can be substan-tially different.

Most owners, contractors and subcontractors are now subject to a quality system which requires the company to maintain procedures covering its activities, however such procedures may not be fully compatible with those of the owner/contractor. In order to ensure efficient working it is necessary to set down those common activities necessary to align existing procedures and to establish new procedures at the administrative interfaces.

The lack of any standardization in documentation and working practices means that there is a risk of inconsistency in the procedures. For example payment requirements may in some instances be stated in the conditions of contract, while in others they are included in the compensation section of the contract, alternatively they may be incorporated in a set of coordination procedures bound into the contract or agreed subsequent to the award of the contract.

Procedures can be divided into management, design and site procedures but the complexity and detail of the documentation will depend on the nature of the project and more importantly the requirements of the conditions of contract.

5.32 MANAGEMENT PROCEDURES

Throughout the whole of this chapter, comments have been made regarding a hierarchy of control and reporting being available at various levels to suit the requirements of the many members of the project team and their corporate masters. Emphasis has been placed throughout on the complex nature of engineering projects both in terms of design and management.

The establishment of a single set of procedures which covers the overall administration philosophy is essential for complex multidiscipline projects to work efficiently.

The following subjects will normally need to be coordinated to ensure compatibility between procedures produced by various organizations:

- communications including addresses for correspondence, filing systems, verbal communications and electronic mail;
- engineering matters including requirements for document distribution, registers, drawing and specification format, drawing review and approval, as-built drawings, certification records and operating manuals;
- procurement, material control and subcontracting, including requirements for purchase orders, approval of all tenderers/contractors, layout drawings of warehouses and storage areas, material receipt and control, certificates and traceability and disposal of surplus material;
- personnel control;
- QA/QC requirements including quality plans and procedures, inspection records and dimensional control;
- safety plan and safety manual;
- progress reporting including requirements for reporting, daily, weekly, monthly, and the provision of progress photographs;
- planning and progress control requirements including programmes at various levels of detail, progress control and measurement systems, histograms and progress 'S' curves;
- cost control and reporting requirements including work breakdown structures, code of accounts, monitoring and controlling costs, reporting and forecasting costs and cash flow schedules;
- invoicing and payment including applications, submission, disputed invoices, payment;

- contract instructions/amendments/variations including notification and evaluation;
- insurance claims reporting and evaluation;
- project completion including final accounts, acceptance certificates and final documentation.

5.33 DESIGN PROCEDURES

The design procedures are probably one of the most complex set of procedures on any project. Not only are the design procedures within a single design team complex, the approval cycles and coordination with other parts of the design process, e.g. equipment manufacturers and owners, make it essential that all parties are working to compatible procedures.

Design procedures cover the management of the design and should include the control both of written information in the form of specifications and reports, etc. and of drawn information.

In the initial phases of a project it is essential to put in place the structure of the design procedures that will be suitable for all the phases of the project. However it is not necessary to have a full set of procedures at the start of a project to cover the later elements of the design process, for example the procedures to cover the review and approval of suppliers' drawings, such procedures can be issued at a more appropriate time. It is essential that the procedures are structured to enable them to be incorporated at a later date and developed to meet the needs of the project.

Procedures are normally based on the owner's requirements and developed to meet the requirements of the project. An owner who regularly procures projects will often have detailed corporate standards which must be complied with. These can cover the provision of drawn information in an electronic form suitable for the owner's own system, and at the other extreme down to the thickness of lines on drawings.

As a project develops so will the design procedures and the owner's initial requirements will be expanded by the designer's, contractor's, subcontractors' and suppliers' design procedures. It is essential that these procedures are aligned in order to avoid clashes and to reduce the risk of complicating systems and formats.

One of the key procedures is that covering the status of design information as it is important that the issue of all information is controlled in order that the receiver of the information understands the status of the information and what they can and cannot do with it.

5.34 SITE PROCEDURES

Site procedures are a set of procedures used to control the work carried out on site. Procedures cover a number of subjects, some of which are legally essential to the lawful operation of a site, particularly in the area of health and safety, while others are essential to the efficient working of the site and administration of the contracts.

Site procedures control various activities covering:

- access to the work;
- working hours;
- contractual matters;
- security/confidentiality;
- health and safety;
- public relations;
- meeting and reporting requirements;
- management structures;
- quality assurance;
- environmental constraints (noise, dust, etc.);
- testing and commissioning.

The procedures vary from project to project and within a project from contract to contract.

Site procedures are normally based on the owner's requirements, which will define a series of restrictions and administrative requirements for the project. These are developed by the design team and may be such that they have to be included in the tender documentation to the extent that they affect the tender price. Contractors may later develop and enlarge on these procedures as may be appropriate for their own control purposes.

5.35 PRODUCTION OF SITE PROCEDURES

Site procedures need to be in place before the commencement of any site work or associated off-site work. The procedures should be clear and well coordinated in order to support the efficient running of a contract. If they are, for example, poorly structured and drafted using different styles they tend to lead to confusion and frustration.

In their drafting it is important to coordinate elements, for example the working hours on site might be restricted by various requirements, such as the owner supervises all work during set hours, noise restrictions at certain times of day laid down in an earlier planning approval, availability of the owner's site security personnel to permit access to the site, etc.

It is not uncommon for a contractor to issue site procedures to the project manager for comment or approval. This should help to ensure that the site will be properly controlled and that any restrictions, etc. are clearly notified to the site team.

The relevant site procedures should be issued under QA procedures as a controlled document to the owner, project manager and contractors and through the contractors to their subcontractors, etc.

5.36 TYPICAL SITE PROCEDURES

The following is a limited list of issues that site procedures cover.

5.36.1 Health and safety

Health and safety is an important element in site procedures and can typically include permits to work, site evacuation etc.

5.36.2 Instructions

The issue of giving instructions is a common contractual matter. Site administration procedures should look to standardize the contractual position and cover matters which are not specifically covered by the contract.

Typically contracts require that all instructions shall be in writing. But what happens if the contractor receives a verbal instruction? Should he act on it? It is common in this instance for the contractor to confirm the verbal instruction in writing and ask for a formal instruction in writing. It must be clear either within the contract or in site procedures whether the original verbal instruction must be acted upon or not before any written confirmation is received and whether failure to reject the verbal instruction means that the instruction is confirmed. Similar issues arise over the status of site meeting minutes and correspondence to the contractor from the owner, etc.

In addition to the issuing of instructions most contractual requirements state that only named persons may authorize an instruction.

Site procedures must be in place in order to actively control the above.

5.36.3 Drawings and specifications

Drawings and specifications are obviously essential for construction to be carried out. This information must be accurate and current; hence good site administration procedures are essential in order to ensure that this is achieved.

It is usual for site administration to have registers of drawings and specifications including lists of superseded information. The control of this information can be very complex when both the owner and the contractor have elements of design responsibility. The owner's design may change and this change needs controlling and communicating. Contractor's design often requires the owner's approval, which is normally required within a specified time period. Any failure by the owner to comment upon or approve the document, may in certain circumstances be taken as approval and any belated comments construed as variations. The situation becomes more complex when the owner's changes affect the contractor's design and in turn suppliers' and subcontractors' design. All this must be controlled and recorded in order that the work is accurately and efficiently carried out while ensuring any performance criteria, required by the owner for the works, are met.

5.36.4 Site records

Site records are numerous and are commonly kept by the contractor with copies being sent to the contract administration team. These may cover:

- weather conditions – temperature minimum, maximum, wind, rain, etc.;
- resources on site – list of management personnel and operatives including discipline and name;
- plant on site – type, duty/size;
- health and safety records – hours worked, incidents, etc.

Each of the above may be recorded at different times varying from hourly to monthly. Some records may be available for inspection by the owner, some copied to the owner and others agreed with the owner.

It is important that the site records requirements are clearly identified and in place as it is common to refer to them at times of contractual or quality disputes.

5.36.5 Payments

As money is the life blood of all companies it is essential to lay down the procedure by which payments are agreed and made. The procedures should be based on the requirements of the contract and should cover:

- timing – monthly, towards the end of the month or event based;
- level of support detail required to be submitted with the payment application;
- person and address to whom the payment application is to be submitted, and any requirement to obtain a receipt on submitting it;
- monitoring the progress of the payment including applications for interest on late payment.

As cash flow is the life line of a contractor's business this is often viewed as a key procedure for all parties to a contract.

5.36.6 Insurances

Although the checking for provision of insurances by contractors/subcontractors may be the subject of a management procedure, site procedures are usually restricted to the manner in which insurance claims should be handled.

Provision of insurances may include obtaining insurance in joint names, vetting of the policies by the owner, displaying insurance certificates, annual renewal of policies, complying with the policy requirements to provide information to the insurers.

Claims cover serving notice of a claim, keeping records, agreeing and progressing any remedial work, obtaining payment, etc.

The subject is a specialist area and it is not uncommon to have the owner's and contractor's experts put this site procedure in place.

5.36.7 Security procedures

Security procedures are often based on the owner's existing requirements which the contractor may develop to suit his needs. Subjects such as the issue of passes, vetting of personnel and the powers to search are common.

5.37 PEOPLE AND PROCEDURES

All good procedures should clearly define what people are required to do under the procedure. They often cover the following:

- giving of authority to a named person or persons;
- giving authority to a person fulfilling a particular job title;
- defining the disciplines/trades of those who can carry out the procedure;
- defining the various duties of those mentioned in the procedure.

Structures are normally prepared which show the reporting lines and duties for a particular procedure. The structure should cover job title, name of persons who carry out the job, telephone number and location. Often they include the names of alternative people who are authorized to carry out the tasks. In certain circumstances these structures might cover more that one procedure when they cover a similar topic.

It is not uncommon to have the same individual carry out various jobs within different structures, but care must be taken to ensure that there is no conflict of interest between the person's roles. For example, the person who is responsible for achieving progress of the work should not be responsible for health and safety.

Having the right structure in place which is not overcomplicated and which includes appropriately experienced and trained people will support the smooth running of a project.

Due to the relatively short-term nature of projects and the changing activities carried out in the duration of a project it is common for the structures to need regular updates.

5.38 CHANGES TO PROCEDURES

It is not uncommon for procedures to have to be changed during a project. The changes vary from minor ones which can be predetermined (e.g. a change in the contractor's site construction team due to the planned retirement of an individual) to major and unforeseeable changes (e.g. the local authority suddenly imposing both weight restrictions on vehicles using certain local access roads and the hours that heavy good vehicles can use the roads). These changes should be managed correctly in accordance with the established QA procedures.

When a change is being considered it must be notified firstly as a potential change and its financial and time implications considered and controlled.

5.39 COMPLIANCE

It is a fact that non-compliance of procedures does happen even on the best run projects. On the discovery of the non-compliance it will often be necessary to notify any defect and reinforce the procedure including providing additional training, etc.

The results of non-compliance can include:

- dismissal of staff (for smoking);
- suspension of the work (on health and safety grounds);
- the contractor being removed from future tender lists (failure to correctly implement site procedures);
- reduced payments (due to not serving notices to the correct time scale).

5.40 CONCLUSION

Procedures are essential for numerous reason but they must be clear and relevant in order that they support the project and do not restrict it with bureaucracy.

BIBLIOGRAPHY

Association of Cost Engineers (1994) *Standard Code of Accounts*, The Association of Cost Engineers, Sandbach, Cheshire.

Clark, F.D. and Lorenzoni, A.B. (1978) *Applied Cost Engineering*, Dekker, New York.

Gray, C.F. (1981)*Essentials of Project Management*, Petrocelli.

Hewerdine, S. (1994) *Plant Integrity Assessment*, Institution of Chemical Engineers.

Institution of Chemical Engineers and The Association of Cost Engineers (1988) *A Guide to Capital Cost Estimating* 3rd edn, The Institution of Chemical Engineers, Rugby.

Jackson, M. J. (1986)*Computers in Construction Planning and Control*, Allen & Unwin.

Karger, D.W. (1991)*Strategic Planning and Management*, Dekker.

Kharbanda, O.P. and Stallworthy, E.A. (1985) *Effective Project Cost Control*, The Institution of Chemical Engineers, Rugby.

Smith, N.J. (ed) (1995) *Engineering Project Management*, Blackwell Science, Oxford.

Stukhart, G. (1995) *Construction Materials Management*, Dekker.

Quality Assurance

6.1 INTRODUCTION

In the current business world, it has become increasingly necessary to obtain a recognized and accepted standard of measured quality before major customers will consider doing business with a company. This fairly recent quality revolution has gained pace and credibility from the early days of the BSI Kitemark, through BS 5750 to today's ISO 9000 series and is now the subject of internationally agreed standards covering goods and services.

The need for a quality system standard was recognized and developed during the Second World War and subsequently in the US space programme. It was apparent that products and systems purchased from contractors would not be available on time or be compatible and work to specification unless each operated according to a strongly defined and documented quality system.

In Europe the requirements of a suitable system applicable to defence contractors were standardized through NATO as Allied Quality Assurance Publications (AQAPs), and were initially published in the UK by the Ministry of Defence as defence standards (DEF Stans.). These standards were subsequently used by the British Standards Institute as a basis for the introduction in 1979 of the BS 5750 quality system standards.

This process of development culminated in the publication of ISO 9000 standards by the International Standards Organization, ISO, in 1987. On their publication the original British standards were withdrawn and replaced by the BS 5750: 1987 series as the UK version of the ISO 9000 documents.

Quality management applies to all products and services — fields as diverse as production engineering, food, textiles, building and construction, energy, health and social services, transport, pharmaceuticals, local government, accountancy and law. Those responsible for quality in all these fields, apply scientific and technical skills and specialized management techniques to achieve the standard of product or service required by the customer at the right price and the right time.

All professionals, whether accountants, architects, engineers, surveyors or solicitors, now have the opportunity to demonstrate their efficiency and consistency through the application of a proven quality management system within their own businesses.

6.2 QUALITY ASSURANCE

There are many 'definitions' of these two words, which being subjective often lead to confusion at first encounter. However, as the reader gains more confidence in the subject, a pattern of similar interpretations will emerge.

The most quoted definition is that given in BS 4778:

> All those planned and systematic actions necessary to provide adequate confidence that a product or service will satisfy given requirements for quality.

As each contract for service or product may vary from the previous one, the perception of quality can readily be seen to rest with the customer. If each customer is satisfied, then quality has been achieved.

So that customer satisfaction is not a 'one-off' or random event, it follows that a systematic approach is required. The rapid growth of a particular system, such as BS 5750 (now renamed ISO 9000) is due to the fact that it offers a framework of activities that assess and measure the installed quality system, and provides for it to be recorded and remeasured for consistency and checks on corrective actions and (it is to be hoped) improvements.

It is quality records, particularly in the form of audit reports, that permit management to review the effectiveness of their business practices, and which give management and their customers the knowledge that their products or services are 'quality assured'.

6.3 TOTAL QUALITY MANAGEMENT

Total quality management is something the majority of companies are pursuing today in one form or another. A caution is, as Professor Steven Wearne (UMIST) said, 'Quality can lose purpose, become a paper exercise, oppressive and bureaucratic'. Every opportunity to prevent such a happening and to build a quality culture is needed.

A new UK standard, BS 7850, which takes more account of total quality management is now available. The objective of a total quality management system is to identify all those matters necessary to ensure the achievement of the written objectives.

6.4 THE STANDARDS

The model for a system of quality management is ISO 9000 that can be adopted by companies and independently evaluated. It effectively combines the disciplines and procedures that make the achievement of consistent quality possible. The standard also focuses attention on the quality system, gives more effective documentation and, when properly understood and applied, makes quality improvement inevitable. This makes the implemented system more effective in reducing errors, hence saving money and improving customer satisfaction.

6.4.1 ISO 9000 and ISO 9004

ISO 9000-1 provides guidelines for the selection and use of the standards and explains the structure of the standards. It defines the quality concepts and clarifies the distinctions and interrelationships between them. Guidelines are also provided for the selection and use of the quality system standard that can be used for internal quality management or external quality assurance purposes. The importance of this section of the standard is that it explains when and how to use ISO 9001, 9002 or 9003 when there is a contract between a purchaser and a supplier. ISO 9004-1 is a *Guide to Quality Management and Quality System Elements* which is designed as an aid to the introduction of internal quality management systems.

6.4.2 ISO 9001

ISO 9001 is the *Specification for Design/Development, Production, Installation and Servicing*. It applies to those companies and organizations that are involved in the original design, development, production and subsequent operation of a product or service as specified by the purchaser. Suppliers covered by ISO 9001 include design groups and computer installers.

6.4.3 ISO 9002

ISO 9002 is the *Specification for Production and Installation*. It applies to companies or organizations providing an established product or service to an agreed specification. The supplier is here assumed to have no significant part in new design or in the servicing, which in many cases are under the control of the purchaser. This leads to more limited responsibilities for management, internal auditing and training. Suppliers covered include manufacturers and suppliers of pressure vessels and valves.

6.4.4 ISO 9003

ISO 9003 is the *Model for Quality Assurance in Final Inspection and Test*. It applies to companies where conformance of the product to specified requirements is required to be confidently demonstrated by inspection and tests on the finished product.

6.5 THE SYSTEMS

The introduction of ISO 9000 quality systems alone does not guarantee product or service quality. This has to be accomplished through a combination of good systems and technical competence. ISO 9000 sets a framework for the journey of continuous improvement towards total business excellence.

6.5.1 Quality plans

Quality planning is an essential part of any quality system and the quality plan describes how and when the quality requirements will be implemented.

6.5.2 Typical contents list of a quality standards system

Most quality systems address the following 20 topics. Some are self-explanatory, all are to be found in ISO 9000 and in most general texts on the subject of quality:

1. management responsibility and organization;
2. quality system or programme;
3. contract review;
4. design control;
5. document control;
6. purchasing;
7. purchaser supplied items;
8. product identification and traceability;
9. process control;
10. inspection (testing and surveillance);
11. inspection (measuring and test equipment);
12. inspection (test and operating status);
13. control of non-conforming products;
14. corrective action;
15. handling, storage, packing and delivery;
16. quality records;
17. quality audits;
18. training;
19. servicing;
20. statistical techniques.

ISO 9000 provides the blueprint for quality systems. It is expressed in terms that can be applied to all types of large, medium and small enterprises, in designing and constructing the product or providing the service to the requirements of the purchaser.

The current international (ISO) standard is the ISO 9000 series, and the well-known British quality standard, BS 5750, was renamed as BS EN ISO 9000 in 1994, to align with the numbering for easy international recognition. The text was and is identical, and future revisions will be similarly in step.

6.5.3 Quality manual

The highest level of the corporate quality plan is the quality manual that formally states the company's policy on quality, and gives a summary of the governing principles in the main areas of the company's business. A typical quality plan will identify a listing of company procedures with a summary description of content to indicate linkages and relationships between the various procedures. This may be shown as a flow chart.

The manual will apply to all operations of the company, and any activities performed outwith the requirements of the quality manual will require prior written approval from those so authorized.

6.5.4 Quality plans

Separate quality plans will be produced and issued for each of the main areas of the company's operations and on major projects the components of the quality plan may be included in the project execution plan. The quality plans will acknowledge the principles laid out in the quality manual and will comprise the procedures agreed and approved by the corporate body relating to specific quality practices, resources and activities relevant to a particular operation. Any departures from the requirements of these procedures must have prior written authorization.

6.6 PROCEDURES

Modern management thinking is that work processes need to be understood and specifically engineered to aid the achievement of objectives.

Records on how decisions are reached are thought by some people to be a problem. Trying to find out what has been considered and what has not, and the reasons for a particular selection, can prove difficult; but an understanding of the principles followed in reaching a decision is essential to validating a quality design and later to operational decision making.

The quality assurance team's objective should be to determine the optimum number of procedures. Procedures have a tendency to go much further than is strictly compulsory on the basis that such further detail aids efficiency. In many cases such prescription is counter to efficiency, is oppressive and limits initiative and improvement.

Procedures should be simple, include only essential material and should only be compulsory for those matters which, if a certain methodology were not followed, would pose a real risk of failing to meet purchasers' requirements, be a risk to safe working, cause financial loss or breach legal requirements. It should be made clear in the procedure what is compulsory and what is merely guidance and therefore not compulsory.

According to the principle of 'ownership', draft procedures should be reviewed before final approval and issue by the persons who will be responsible for implementing them.

Below the level of formal procedures there will be job instructions or method statements that define the details of how the procedures are to be implemented. Such documents should be produced by the person responsible for the function, who must also approve any significant departures from their requirements.

6.7 TEAM QUALITY

Although the various managers on a project are responsible for the implementation and maintenance of quality systems within their areas of responsibility the responsibility for quality rests with everyone on a project.

It is common to appoint a QA engineer to act as a focal point and catalyst for QA matters who will normally be responsible for the following:

- providing a service and advice on quality matters;
- ensuring the implementation of necessary quality systems throughout the project and any matters not specifically assigned in this section;

- arranging and where appropriate, conducting quality audits through-out the project;
- coordinating and standardizing all supplier surveillance activities.

Contractors have in many cases instigated quality councils and training to involve everyone.

The quality assurance team should aim to 'engineer' the work or decision processes used, rather than evolve 'a sticky plaster approach' to putting problems right as they occur.

6.8 SUBCONTRACTOR AND SUPPLIER QUALITY

It is now common to rely on the quality assurance of the subcontractor or supplier rather than on the traditional approach of the purchaser inspecting the work. In certain cases (a common example being British Steel), the importance of the material is such that independent inspection is still required. In other cases though it should be satisfactory to validate that the supplier has adequate inspection and rely on that.

Suppliers should be registered to ISO 9001 or ISO 9002, or have equivalent procedures in place.

A team comprising senior managers will often visit all important sub-contractors and suppliers prior to or shortly after award of a contract. This is to impress on the subcontractor or supplier the importance of their work to the project, the need for quality and timely completion.

Quality assurance programmes will be developed for all procured items as part of preliminary engineering.

A performance test will be performed wherever practicable on all suppliers' equipment.

Suppliers will generally be required to maintain material and test records for inspection throughout the operating life of the project.

6.9 QUALITY IMPROVEMENT

There are always two bases for quality improvement: one is the reactive basis, where problems discovered during audits or other surveillance methods are recorded, and preventive measures taken to ensure no recurrence. The second is the proactive, which is more difficult as improvement implies knowing the starting point. There is, therefore, a need always to consult data collected with the intent of analysis for quality improvement, or data that are available that can be used for quality improvement analysis, the proactive being less well defined.

The main tool for recognizing, in a proactive way, the need for quality improvement, is the production, where possible, of flow diagrams for each working procedure. The logical sequence can be seen and, therefore, the expected outcome can be assessed.

6.10 THE WAY FORWARD

The future success of all our businesses is in our hands, to win work, to complete it to the customer's satisfaction, to encourage further orders and to ensure continuing improvement in business methods. In this search for improvement the three 'soft' management issues of quality, safety and environment will rank equally in importance to the traditional 'hard' issues of cost and schedule.

It is self-evident that a quality job will be a safe job, and noting that BS 7750 for environmental systems is based on the old BS 5750, we have the basis for considerable synergy by combining the three subjects into one structure. Purchasers cannot abdicate their legal responsibilities to the contractor; the law now makes the 'employer' responsible. Future work must be won by competing in all the above areas, and it will be the most efficient team that wins the business.

BIBLIOGRAPHY

A quick check on the status of quality systems in place in most formally assessed companies can be found in the Department of Trade and Industry Register.

Ashford, J.L. (1989) *The Management of Quality in Construction*, E & F N Spon.
Department of Trade and Industry *Guide to Registered Firms*.
Lock, D. (1995) *Handbook of Quality Management*, Gower.
Owen, F. (1993) *Quality Assurance: Guide to Application of ISO 9000*, Institution of Chemical Engineers.
Parker, G.W. (1995) *Achieving Cost Effective Quality*, Gower.
Sayle, A.J. (1995) *Management Audits*, McGraw-Hill.

The Contract

Isomerisation Unit at
Elf / Murco / Gulf Refinery,
Milford Haven, Wales,
showing the
deisohexanizer.

7.1 INTRODUCTION

A contract may comprise a number of documents usually bound together, by reference, in a form of agreement signed by the contracting parties. A construction contract will therefore typically be comprised of the following in one form or another:

- form of agreement;
- conditions of contract;
- supplementary or special conditions;
- scope of work;
- compensation or terms of payment;
- coordination procedure;
- technical information including drawings, specifications;
- programme requirements, etc.

The form of agreement will not only list the above, or similar sections making up the contract, but may also state the order of precedence in the event of a contradiction between the documents. Other aspects which may require consideration according to the circumstances of the project are listed in section 7.18 of this chapter under supplementary and special conditions.

The form of agreement is often part of a set of conditions of contract, which sets out the responsibilities of the various parties and contains the more standard procedures under which the contract is to be administered. Procedural information, in addition to any contained in the conditions, is discussed in Part 5 of Chapter 5.

The form of agreement and conditions of contract are frequently referred to as the form of contract and may be based on, or take the form of, a standard printed document. A standard form of contract, which has been formulated and authorized by one of the professional institutions, is advantageous where it can be directly related to the type of work concerned. In many cases the wording of the clauses contained in the standard form of contract will have been tried and tested over a period of time and may even have been tested in courts of law.

In practice many engineering contracts are based on non-standard forms of contract; in such circumstances it would be advisable to compare their various provisions with those advocated in standard forms in order to anticipate their effect. Many of the larger companies, such as chemical and oil companies, have their own 'standard' form of contract developed over the years, as it would be difficult for them to use published standard forms for work such as offshore construction, without extensive amendment.

No two projects are ever the same and, where the conditions are based on a standard form, clauses may be amended and additional clauses introduced to suit particular circumstances. These must be read carefully in order to ascertain their full effect, particularly as they relate to the obligations and responsibilities of the parties.

It is most important that a contractor fully recognizes the responsibilities he is expected to accept when tendering. These responsibilities must be set out unambiguously in the contract. Most standard forms attempt to establish an equitable relationship between the parties but some owners may choose to transfer the onus of certain obligations on to the contractor. The increase in the risks to be undertaken may affect the financial provision required by the contractor and included in the tender price.

7.2 TYPES OF CONTRACT

The choice of the type of contract will inevitably vary according to the owner's requirements and the nature of the project. The more usual types of contract tend to fall into the following categories which are further described in Chapter 9 on contractor and subcontractor selection:

- turnkey based on lump sum;
- lump sum: design and manage;
- remeasurement based on schedule of rates /approximate quantities;
- performance contract: design and construct;
- cost reimbursable (which can be subject to a target cost).

The above list can only be indicative of the many types of contract in general use. Many contracts use a combination of contract types: lump sums for elements of work for which the scope can be ascertained, while approximate quantities may be appropriate for other elements the scope of which cannot readily be determined at the time of tender.

Types of contract will vary considerably, depending on the responsibilities to be borne by the parties and the extent to which the scope and definition of the work can be established prior to tenders being invited for the work, or the tender accepted. The type of contract may also vary between the owner and the contractor and the contractor and the various subcontractors. The selection of the type of contract is discussed in Chapter 9.

Arrangements between the parties will recognize the risks involved, which may lead to different contracting methods and the means whereby these risks can be shared. Such arrangements include joint venture contracts, where two or more contractors agree to collaborate in submitting a tender, and alliancing, where various contractors and an owner agree to work together in a cooperative agreement and share risk and savings.

7.3 FORMS OF CONTRACT

The following is an indicative list of the more usual standard forms of contract used in the industrial engineering industry, which can be regarded as having been standardized and accepted by the industry:

- the Institution of Chemical Engineers: *Model Form of Contract for Process Plants: Lump Sum Contracts* ('Red Book');
- the Institution of Chemical Engineers: *Model Form of Conditions of Contract for Process Plant: Reimbursable Contracts* ('Green Book');
- the Institution of Chemical Engineers: *Model Form of Conditions of Contract for Process Plant: Subcontracts* ('Yellow Book');
- the Institution of Civil Engineers (ICE): *Conditions of Contract*;
- the Institution of Civil Engineers (ICE): *Design and Construct Conditions of Contract*;
- Fédération Internationale des Ingénieurs-Conseils (FIDIC): *Conditions of Contract for Works of Civil Engineering Construction*;
- Fédération Internationale des Ingénieurs-Conseils (FIDIC): *Conditions of Contract International for Electrical and Mechanical Works*;
- the Institution of Electrical Engineers: *Model Forms of General Conditions of Contract: Engineering Contracts* (MF/1).

One form of contract which has been well publicized but which has still to be fully tested is the New Engineering Contract: the engineering and construction contract issued under the auspices of the Institution of Civil Engineers.

7.4 TERMS AND CONDITIONS

The terms and conditions set out in the form of contract, whether they form part of a standard form or have been drawn up specifically for a particular project, are often required to cover a wide range of circumstances. Some will be peculiar to a specific industry or to a particular type of project.

The principal aspects which need close examination as to the adequacy, implications and onus of the respective parties and are common to most contracts are set out in this chapter and comprise:

- definitions and interpretations;
- scope of work;
- responsibilities;
- law and statute;
- damage, injury and insurance;
- subcontracting;
- time;
- variations (changes);
- payment;
- testing, take-over and liability for defects;
- default and remedies;
- suspension and termination;
- resolution of disputes.

7.5 DEFINITIONS AND INTERPRETATIONS

Terms vary from one contract to another and between standard forms. It is therefore important to ensure that all the contract documents comprising the contract use the same terms and definitions throughout. The terms in the conditions of contract should be defined clearly to avoid any ambiguity in their interpretation and adopted and used throughout the contract documentation and subcontract documents.

The clarity of these definitions will assist in eliminating possible misinterpretation of the contract arrangements and procedures, thus contributing to the smooth running of the project.

7.6 SCOPE OF WORK

One of the most important definitions required in a contract is that of the scope of the work. This must be carefully drafted in order to define clearly what is expected of the contractor in the performance of the contract and therefore included in the quoted tender amount.

It must be remembered that the scope of work is one of the fundamentals of the contract. If the contract does not cover all aspects of the work, either specifically or by implication, that aspect may be deemed to be excluded from the contract. A comprehensive, all-embracing description should therefore be considered for the scope of work clause which should be supplemented by specific detailed requirements. If reliance is placed solely on a very detailed scope description an item may be missed from this detailed description and be the subject of later contention.

Where items of equipment are to be fabricated or manufactured off site by others, it is advisable that the contract sets out the corresponding obligations and liabilities of the respective parties, particularly if these are to form an integral or key part of the completed works.

7.7 RESPONSIBILITIES

The responsibilities of the contrasting parties and their representatives should be clearly defined in a manner which will leave no doubt as to the obligations that each is accepting. This will also allow the procedures necessary to enable the contract to progress satisfactorily from inception to completion to be established at the outset of the work.

The person given ultimate responsibility on a project may be entitled project manager, owner's representative, engineer or supervising officer (in this chapter this list of titles is abbreviated to 'the engineer'). The engineer will be identified and named in the contract and, as an agent of the owner, will represent him by interpreting his requirements and will issue appropriate instructions on the owner's behalf to the contractor. Any limitations to the powers of this person when acting for the owner will need to be clearly stated.

Where the engineer delegates any part of his responsibilities, the extent of such delegation must be defined. In this way, the engineer's representative, cost engineer or quantity surveyor, etc. will each have their respective authority, duties and obligations defined for the purpose of the contract.

The contract as a whole will also define the responsibilities for the design, production and programming of the design information, the requirements for design approval, the supply of any free issue materials, the issuing of instructions and the form which these instructions are to take, the programming of the works, the method of measuring and evaluating the work, the circumstances which will constitute a variation to the scope of works and the duties of the parties during construction and installation.

7.8 LAW AND STATUTE

The contract should specify the law under which it is to be interpreted as this may not be the law of the country in which the work is to be executed.

Differences in the law between countries need to be carefully considered. Local laws may be in place which affect health and safety, employment matters, the manner in which the work is carried out, how disputes are resolved, the ownership of plant, materials and equipment prior to their delivery to site, patent rights, copyright, royalties and their equivalent, etc.

7.9 DAMAGE, INJURY AND INSURANCE

Generally, standard forms of contract deal separately with the responsibility for loss or damage to the works and for injury to persons and loss or damage to other property.

Under most standard forms of contract used in the construction of engineering projects, the contractor will be made responsible for the care of the works and their restoration if loss or damage occurs before they are certified as completed. Separate clauses deal with the insurance requirements to cover this liability. Exceptions to this responsibility will be found, for such risks as war, *force majeure*, etc.

The obligations to insure the works, as distinct from the employer's liability and third party insurance, may well depend upon the size of the contract and its form. The contractor may be required to insure the works under an engineering contract, but the owner may insure for the benefit of all the contractors and himself where the contract is of large scale, or where disproportionate risks exist such as an offshore platform, or where a number of contractors are to be engaged on a construction site when it would be difficult to ascertain cause and liability for an incident.

Most contract forms will seek, for the benefit of the owner, an indemnity from the contractor for injury to third parties and for loss or damage to property, other than the works, which arise out of or as a result of the construction work. That indemnity may well include the acts of parties for whom the contractor is responsible as well as those of his employees. The owner will also require the contractor to insure to cover this indemnity.

In fulfilling his insurance obligation the contractor will be required, normally at the commencement of the contract, to substantiate the fact that insurance has been arranged.

Reference to responsibility for injury, loss or damage and the requirements to insure is provided in Chapter 8.

7.10 SUBCONTRACTING

Most contractors have found it uneconomic to retain the full range of specialist expertise necessary to undertake all aspects of a multidiscipline project. The contractor may therefore need to subcontract some of the work to other contractors because of the specialist nature of parts of the work, equipment, engineering or plant to be incorporated.

The owner may require to be involved in the selection and appointment of these subcontractors and may require particular subcontractor responsibilities or design parameters to be incorporated in appropriate documentation. These arrangements may involve nomination, designation, restriction as to selection or other requirements which may involve close cooperation between the owner and the contractor either prior to or during the contract period.

In many cases, documentation drawn up by the contractor for the appropriate subcontract will reflect the provisions of the main contract between the owner and the contractor.

7.11 TIME

Delay and disruption to the work is the most frequent cause of claims being raised by contractors against owners, while delay in completion due to failure by a contractor is the most frequent cause of action being taken by owners against contractors.

A failure in performance of a responsibility for design by either party, may lead to delay in providing the information necessary to enable the works to proceed. Alternatively the contractor may fail to meet his programme requirements by poor progress, possibly caused by delay in the provision of labour and material or due to his subcontractors or suppliers. Failure to provide design information, free issue material, plant or specialist equipment promptly and correctly is one of the most common causes for delay, with accompanying financial penalties under the contract. This may be compounded by errors or omissions in the drawings or other design information.

Circumstances outside the control of either party such as inclement weather, national strikes, hostilities, unforeseen ground conditions, may also prevent the programme from progressing as initially envisaged.

The proper control of these events is necessary if the contract is not to be frustrated and unnecessary claims submitted. Agreed start and completion dates are essential and, depending on the size and complexity of the project, an agreed contract programme may also be either essential or desirable.

The conditions of contract will contain clauses which recognize the various circumstances which can adversely affect or vary the work and the programme. Failure to recognize a change in circumstances and deal with it equitably and expeditiously can lead to friction between the parties and possible litigation and arbitration.

Some contracts, where the contractor is responsible for the whole of the project, will stipulate only the date on which the contractor is required to commence the work and the date by which the work is required to be completed and ready for handover by the contractor to the owner. The details of how the work is to be organized are then left to the contractor to establish and monitor, although he may be required to submit his detailed programme in advance for approval.

In other contracts a detailed programme, rather than just the start and finish dates, will be regarded as a contract document and may require the owner's agreement to any alteration. This is often the case where a number of 'main' contractors for civil, mechanical, electrical, instrumentation and insulation, are involved and key dates have to be established to ensure site and work availability for each.

In either case a programme will be required to show the contractor's proposed method and sequence for the execution of the works. The programme may also indicate the proposed sequence and dates for receipt of key documents and information to be provided by the owner together with similar details of any plant or specialist equipment for which the owner is responsible for procurement.

Where materials and equipment are to be provided on a free issue basis, dates for delivery should be established by the owner and the contractor advised accordingly. Failure by the owner to meet these dates may cause delay and incur financial penalty.

In support of the programme the contractor may be required to provide detailed schedules for individual sections of the work and progress charts and manning levels for approval by the owner and may subsequently be required to update progress charts during the progress of the contract.

In most cases it is the engineer who is conversant with the development and progress of the work and who will best be able to communicate in writing to the contractor those decisions which will determine the smooth running or otherwise of the contract.

7.12 VARIATIONS

A variation is a change or an alteration to the work which involves an addition, omission or substitution to the work described and established at contract award, or a change to the circumstances, including programme requirements, under which the work was to be undertaken and may be variously referred to as engineer's instructions, variation orders, change orders or other such term.

Variations should be in writing. If the variation is in the form of drawn information it should be issued in writing, while an oral instruction given in meetings or at site should be confirmed in writing.

The most suitable vehicle for the evaluation of variations is that provided by a schedule of rates or bills of quantities prepared in accordance with a set of measurement principles such as the Standard Method of Measurement for Industrial Engineering Construction.

In certain cases, because the full extent of the work cannot be predetermined, the only means of valuing the work will be by recording actual time expended on a daywork basis. The record of labour, plant and materials used will need to be signed by an authorized representative on site at the time the work is carried out. Rates and percentage additions established at tender stage may then be applied for overhead, profit, etc., to enable the variation to be evaluated.

The evaluation of a variation may be carried out by the engineer, his representative, or by the contractor. The evaluation may involve a simple calculation of revised quantities against original tendered rates, or possibly an evaluation on a daywork time and cost basis. Where applicable the evaluation of the cost of disruption to the programme caused by the variation or the cost of extending the contract period must be added.

On certain types of contract, the whole of the work will be remeasured on completion, in which case the contractor may be responsible for preparing and submitting priced detailed measurements of the completed work. Alternatively the engineer may be responsible for these duties or appoint a third party to carry them out. The remeasurement may have to be submitted on completion of the contract or on a continuous basis in support of interim applications for payment throughout the course of the works. In such circumstances individual variations would not be remeasured in detail, but would have an estimate attached at time of issue for purposes of cost control. Either way the procedures for authorization and evaluation should be established at award.

Major variations, or variations introduced late in the programme, may cause a disruptive effect sufficient to delay the programme or require special arrangements to be introduced to offset the effect. In many cases it is the combination of a series of variations which accumulates into a significant problem. A delay affecting the work of one of the earlier trades may set off a chain reaction which affects each of the subsequent trades or contracts.

Where an element of the work has not been defined at tender stage, to a sufficient degree to allow the tenderer to include the work in his tender, a specific contingency may be included at the time of the enquiry by the owner in the form of a provisional sum.

Where provisional sums are included in the contract price, these must be replaced at final account stage with a properly quantified and priced set of items or agreed lump sums substantiated in an appropriate manner.

7.13 PAYMENT

The contract should include terms of payment detailing the procedures for interim payment, and retention and should state how the final account is to be calculated and final payment made.

It is usual for the contractor to submit an application at monthly intervals to the engineer indicating the value of the work completed and any other amount such as establishment charges to which he considers he is entitled under the contract. In the case of construction contracts this may include the value of unfixed materials and plant both on and off site. Supporting documentation may be required in certain instances, the details of which will depend on the nature of the work.

The contract normally requires the engineer or his representative to certify the value to which the contractor is entitled. Interim certificates will be issued by the engineer authorizing payment from the owner within whatever period is stipulated within the contract. Delay in receipt of payment to the contractor may entitle him to interest charges. Alternatively the contract may specify predetermined stage payments rather than monthly interim payments.

Where there is an amount to be retained against satisfactory completion of the work, the percentages to be applied for such retention and the conditions and time of release should be clearly stated in the contract.

The maximum period within which agreement is required between the parties as to the precise value of a final account or the final contract price may be stated.

7.14 TESTING, TAKE-OVER AND LIABILITY FOR DEFECTS

Various liabilities commence or are released when certain stages have been reached in the work. Such stages are usually acknowledged by the issue of a certificate, which may include certificates of handover and take-over, partial or final completion, maintenance period.

Industrial engineering projects, by their very nature, will require tests to be conducted both during and on completion of individual installations and in many cases in relation to a number of separate installations which operate in conjunction with each other to meet the design parameters of the project as a whole.

Testing therefore may relate to individual material components, factory testing of completed items of equipment off site, pressure testing of individual sections of pipework, testing of welds or pipe joints, a vast array of electrical and process control tests and tests of environmental conditions or quality of finished product after processing. Testing may also be required for insurance purposes.

Testing needs to be interpreted separately from pre-commissioning or commissioning, as does performance testing where the whole or part of an installation requires certain preset specified design parameters to be met. A more comprehensive description of pre-commissioning and commissioning is given under initial operations in Chapter 2 on the management of engineering projects. The owner may arrange for commissioning of completed installations or groups of installations to be carried out independently or jointly with the contractor(s).

While the specification will set out the procedures and parameters for testing and commissioning, the contract will set out the obligations as to how these parameters are to be met and the consequences of failure to meet them. The responsibility for the provision of fuel for testing and testing equipment also needs to be stipulated.

Performance testing is usually applicable where the design is the responsibility of the contractor and can include the running of the completed plant and the checking of product quality and quantity, feedstock consumed, use of energy, waste and by-products, environmental conditions and other aspects such as may be required. In such instances time limits for the rectification of performance defects must be stated together with penalties for failure.

The documentation required in the form of test certificates, warranties and guarantees, to enable proper records to be maintained should be stipulated, as will the effect in relation to the guarantee or warranty period and the extent to which it will apply.

In many cases the awarding of a handover certificate following testing will establish the date of take-over of the installation or, where applicable, that part ready for commissioning.

Liability for defects should be defined in the contract together with the period of time for which such liability is to apply and the length of time within which defects are to be remedied.

7.15 DEFAULT AND REMEDIES

Where the contractor has failed to comply with the terms of the contract, the owner may have the right to take appropriate action to remedy the default including, where applicable, having the work completed by another contractor at the expense of the original contractor.

The manner of serving notice of default and the financial consequences of the default should be stated.

Where either party to the contract has defaulted to the extent that financial redress or litigation is invoked, contemporary records will be required by way of substantiation. Proper record taking procedures should already have been established but, as soon as potential default situations become evident, these procedures may need to be reviewed and additional records kept.

7.16 SUSPENSION AND TERMINATION

Circumstances may arise where it is necessary for the work to be suspended and the reasons for which such action will be allowed should be stated together with the financial consequences. The consequences will vary according to whether the reason is due to default on the part of the contractor, default on the part of the owner or unexpected happenings such as weather conditions or *force majeure*. The periods of notice should be given for each situation together with the rights and remedies open to each party according to the circumstances.

Termination of the contract rather than suspension may be the only appropriate remedy in the case of insolvency, default, *force majeure* and the like and periods of notice required. Again, periods of notice required and the manner of serving such notice should be stated.

The rights and obligations of each party in the event of termination must be stated, including the right of the owner:

- to engage another contractor to complete the works;
- to use constructional equipment belonging to the original contractor, where the contractor is the party in default;
- to take over possession of all materials and plant, both on and off site;
- to regain ownership of property temporarily in the custody of the contractor.

7.17 RESOLUTION OF DISPUTES

Most standard and bespoke forms of contract will incorporate a condition relating to the way in which disputes are to be resolved. Where the owner and the contractor fail to agree upon any matter covered by the provisions of the contract, the contract may provide that either party may take the matter in dispute to arbitration.

Arbitration provisions may include the periods of notice required to be given by either side, the nomination of an independent body to appoint an arbitrator if the parties fail to agree on such an appointment, the time and venue of any arbitration and the provision of the appropriate legal statutes which would govern the conduct of the arbitration.

The appointment of a third party with a proper knowledge and experience of the construction process and who has the confidence of both parties can often establish a settlement of a particular dispute within a relatively short period of time. In most cases such procedures can be run in parallel with the construction programme without disrupting the work to the benefit of both parties.

Alternative processes for the mediation of disputes rather than arbitration may also be stated in the contract conditions. Alternative dispute resolution (ADR) is a mechanism for resolving disputes which is intended to avoid litigation or arbitration, limit the time and the cost of the action and achieve early agreement.

Sir Michael Latham, in his report entitled *Constructing the Team* recommends that a form of adjudication be incorporated in construction contracts in order to provide an immediate resolution of disputes so that the actual construction may proceed without hold up and limiting the right of appeal, e.g. by way of arbitration, until the works have been completed.

The government in Part II (Construction Contracts) of the Housing Grants, Construction and Regeneration Act has specified that within a 'construction contract' (which term embraces agreements with professional advisers) the parties to the contract shall have a right to refer a dispute for adjudication under a procedure which complies with the Act. Where there is no such adjudication clause in the contract or one which is not in compliance with the adjudication provisions of the Act, then the scheme for construction contracts shall apply. At this time the draft of the scheme is still to be issued by the government.

The requirements under the Act are likely to be very onerous, for example the Act initially requires the adjudicator to reach a decision within 28 days of the referral of the dispute.

As the Act has only just been placed on the statute books it is impossible at this time to provide any advice based on practical experience.

7.18 SUPPLEMENTARY AND SPECIAL CONDITIONS

The previous sections relate to the main areas which are common to most contracts. Individual contracts will inevitably contain clauses required by a particular owner or a particular project, which may be called supplementary or special conditions. This title can be important since in law special conditions take precedence to general conditions, unless otherwise particularly stated to the contrary.

A checklist of the most common clauses is included below. Some, not all, will inevitably be required:

1. services to be provided by the owner;
2. hours of work;
3. labour agreements;
4. special requirements;
5. security;
6. safety, health and welfare;
7. confidentiality;
8. advertising/publicity/photographs;
9. site conditions/hazardous areas;
10. explosives and fires;
11. testing or materials samples;
12. direct contractors;
13. fluctuations;
14. currency changes;
15. nominated or designated subcontractors;
16. liquidated damages;
17. daywork;
18. design;
19. audit requirements;
20. value added tax.

Indemnities and Insurances

8

Lifting a reactor vessel at
Hydrofiner plant
Grangemouth, Scotland.

Plant Owner B.P. Oil Ltd.

8.1 INTRODUCTION

Due simply to the sheer size of the subject it would be inappropriate in this chapter to attempt to provide a detailed and comprehensive recital of the law of contract in relation to indemnities, or advice and guidance on all forms of insurance needing to be considered by an owner and a contractor undertaking construction projects.

This chapter is intended therefore to provide information that will be sufficient to secure a basic understanding and thereafter form the framework for both asking the appropriate questions and making suitable enquiries.

It is also the intention to confine legal comment to the common law of England, although recognizing that work undertaken in Scotland and in other parts of the world may lead to differing legal interpretations.

8.2 GENERAL

It is very common to observe within construction contracts clauses that seek indemnities normally in a format that requires the indemnifying party to provide compensation in circumstances that would not otherwise exist in law. Such clauses are again normally supported by an obligation on the indemnifying party to arrange certain types of insurance to support the indemnity being granted.

Insurance is, however, not confined to supporting an indemnity but extends to operate with other obligations such as the contractor's duty to carry out and complete the construction contract.

The wording or indeed interpretation of such clauses can become complicated when the contractor carries some form of design liability or undertakes a 'turnkey' contract, especially in deciding the extent of the indemnity in relation to the design responsibility and the insurance protection the contractor is required to provide.

The insurance market is very cautious about granting either material loss or liability insurance due to defects in design because of bad experiences in the past. In consequence a full understanding of the extent of that design liability is very important and great care is needed in securing the appropriate cover.

8.3 DEFINITIONS

A study of the indemnity and insurance clauses contained in the various conditions of contract relating to industrial engineering will quickly demonstrate a considerable variation in their wordings.

Many construction projects may include a high value for mechanical and electrical work but yet may still be based on a JCT (Joint Contracts Tribunal) form of contract for building work, an example being the construction of a new trust hospital. ICE (Institution of Civil Engineers) forms of contract utilized mainly for civil engineering projects can again be the base contract for engineering works. Both JCT and ICE forms adopt the terms 'employer' and 'contractor' although representation is by the 'architect' and 'engineer' respectively.

In form MF/1, recommended by the Institutions of Mechanical and Electrical Engineers and the Association of Consulting Engineers and in I.Chem.E., recommended by the Institute of Chemical Engineers we find the 'employer' described as the 'purchaser'.

To avoid confusion in this chapter reference will be confined in the main to calling the 'employer' or 'purchaser' the 'owner', the representative the 'engineer', the construction or engineering work as the 'construction project or contract or work(s)' unless there is a need to identify more specifically.

The words 'peril' or 'perils' are utilized by the insurance market as alternatives to 'risk' and in this chapter are intended to convey a type of risk. Examples of specified perils are the risks of fire, explosion, storm and flood.

The term indemnity has a dual role but the reference to 'indemnity clauses' will relate specifically to the indemnity granted in a construction contract although the term 'indemnity' or its derivatives will also be utilized in relation to compensation paid by insurers. An indemnity clause, dealt with in more detail later in this chapter, is an undertaking by one party to compensate another party for wrong done or trouble or expense or loss incurred. Utilized in connection with insurance policy the term simply means the undertaking of an insurer to compensate the insured for a loss suffered or a liability incurred which arises out of the event or events for which the insured has taken out insurance.

Insurance to support an indemnity clause is mainly confined to claims for bodily injury or physical loss or damage, although cover can be widened to embrace other forms of liability such as obstruction or trespass. However, it is important to remember that all policies are subject to terms, e.g. conditions and exceptions.

8.4 NEED FOR INDEMNITIES AND INSURANCE

Over the years owners have sought the protection of indemnity clauses that not infrequently widen the liability that a contractor would otherwise incur, and have also imposed obligations to insure mainly as an assurance that there will be sufficient funds to ensure that repairs or reinstatement will be carried out and without delay.

Indemnity clauses, normally requiring a contractor to indemnify an owner in defined circumstances, e.g. due to negligence, omission or default of or even breach of statutory duty by that contractor and indeed others for whom the contractor is responsible, need to be studied carefully by a contractor at tender stage. The study necessitates the contractor having sufficient 'internal expertise' or otherwise to seek such expertise, to establish that such clauses have been reasonably drafted and embrace only liability in scope and quantum that the contractor feels can be accepted without jeopardizing the contractor's trading ability. It is important to recognize that an indemnity clause can be drafted to be so wide in scope that it may not be possible to secure insurance that will provide a full protection.

Again, as is explained in section 8.5, there is a need to ensure the insuring obligations are fully understood and are so drafted that they neither introduce ambiguity nor are so wide that compliance is impossible.

8.5 PROVISION OF INSURANCE

Insuring clauses will require the contractor to obtain insurance that normally must be evidenced at the commencement of the contract by the production of certificates or policies of insurance. However, in drafting a construction contract the owner may decide he will undertake the responsibility for arranging certain insurances and likewise will undoubtedly be required to produce evidence of that insurance in the same way. Whichever party does arrange the material loss and liability insurances it is normal for them to be for the benefit of both parties to the contract.

The extent of the insurance can vary but it is general for at least one clause to relate to the insurance of the work being undertaken, including material and goods for incorporation into the work and possible constructional plant. Yet another may specify the scope of the liability insurance required, very often drafted in general terms that are to the effect that the contractor will arrange insurance that compensates the contractor for the legal indemnities the contractor has granted to the owner.

Where such loose descriptions are utilized the contractor may well find that it is difficult or indeed impossible to secure insurance that is fully in line with the scope of the indemnity being granted. This position has existed for many years and appears to have been accepted by owners by default or otherwise possibly in the knowledge that the indemnity clause(s) are paramount.

An example would be ICE (6th edition) where although clause 23 (third party insurance in respect of injury or damage) sets out to be specific, the obligation to insure against 'injury to any person' tends in the more standard form of public liability policy to refer to 'bodily injury' rather than the wider term 'personal injury', there being a considerable difference in law in that personal injury can embrace forms of liability which do not necessitate actual injury to the body.

Where design forms part of the construction contract the indemnity clause may be sufficiently wide as to seek an indemnity for injury loss or damage resulting from design failure, including failure to meet the plant performance requirements as well as defects in materials or workmanship. Nor is there always a specific reference in the corresponding insurance clause in the contract to provide 'professional indemnity insurance', and the public liability policy relied upon may well be limited as to the extent of cover provided in relation to design responsibilities. That limitation is generally in the form of an exclusion whereby an indemnity will not be granted if the injury, loss or damage arises out of a breach of a professional duty, e.g. the responsibility to design without defect.

It is not unreasonable to state that the arranging of insurances for construction projects (in particular large process engineering projects which embrace responsibility for testing and possible commissioning) can be a veritable minefield and that any contractor so undertaking should be satisfied that the necessary 'internal expertise' exists, or otherwise seek the assistance of an insurance broker skilled in the construction industry and its insurance requirements.

Often too little attention is paid at the time of tendering and award to the obligations to insure, let alone to a detailed analysis of the scope of the cover being undertaken. Inevitably this will be to the detriment of the insured parties if later a major claim or loss occurs.

8.6 SPECIALIST ADVICE

In the context of the total income received by the UK insurance market, the construction industry is but a small part and as a result it is unlikely that the majority of staff of intermediaries (e.g. insurance brokers) and insurance companies will have the knowledge required to adequately advise owners and contractors undertaking industrial engineering projects.

It would be sound advice to ensure, if an intermediary is utilized, that the company has considerable experience in the field of construction insurance and a full understanding of the forms of contract utilized in the industry. It is likely that the majority of insurance companies undertaking this type of insurance will have specialist staff capable of assisting the owner, contractor or the intermediary.

8.7 STANDARD CONDITIONS OF CONTRACT

The standard conditions of contract printed in the UK are numerous in themselves and as a number of companies prepare their own conditions of contract reference to all forms of contract in any detail would be quite impossible.

8.8 CONTRACTOR/SUBCONTRACTOR

In the underlying relationship of contractor and subcontractor it is essential for a contractor to recognize that the contract conditions sought from subcontractors must be complementary to the terms of the main contract or, for example, indemnities given by the contractor to the owner may not balance with indemnities received by the contractor from the subcontractor. Insurances arranged by the contractor and sought from a subcontractor may also either duplicate or leave areas of responsibility unprotected if great care is not taken.

It is certainly not unusual to find at the time of an event that a subcontract has not been signed and that the subcontractor's tender was stated to be subject to the subcontractor's own terms and conditions.

It is as well to remember in law a contract needs an offer and an unqualified acceptance. If this 'rule' is not followed great difficulty can arise as to which terms apply and hence which indemnities operate and which insurance should have been provided.

8.9 INDEMNITIES

From the very fact that the undertaking of a construction contract involves an exposure to risk of the contracting parties, there is a tendency in the drafting of construction contract forms to embrace what are commonly known as 'indemnity clauses'.

Clauses of this nature are normally drafted within the main contract for the benefit of the owner and similarly for the benefit of the contractor in subcontract conditions of contract into which the main contractor may enter as permitted or authorized by the owner. It should be mentioned that most standard forms of main contract embrace a clause or clauses dealing with both 'assignment' and 'subcontracting', neither normally being allowed without the express consent of the owner. In addition, such clauses may specify part of the terms that must be embraced within any subcontract conditions.

An indemnity clause is one in which one of the parties to the contract agrees to compensate the other party against certain specified losses in the event of defined happenings occurring. It should be added that in certain standard contracts the clause is so drafted that each party seeks an indemnity from the other in defined circumstances.

The type of indemnity clause included in construction contracts generally relates to loss or damage to property both real and personal, to the death or injury of a person and seeks to secure an indemnity for one party by another which is not simply confined to claims but embraces all actions, demands, costs, charges and expenses.

8.10 DRAFTING OF CLAUSES

Before entering into a contract a study of the precise wording of such indemnity clauses is paramount.

One of the major aspects in the study is to establish, for example, if the owner is seeking an indemnity in circumstances where the owner can be seen to have been negligent. Previous cases heard before the 'courts' have demonstrated the reluctance of the 'courts' to embrace within such indemnities the negligence of the party being indemnified. Because of the dislike of judges to see the transfer of such liability, there appears to be an attempt to recognize such a clause as being 'wholesome' without the importation of that negligence. In other words, unless the wording is so clearly drafted as to be beyond doubt, past case law appears to demonstrate that the judgement will be in favour of the party granting the indemnity, i.e. negligence will not be 'imported'. If the negligence of the owner is to be embraced within the indemnity clause the clause must state so in clear terms.

If the clause is shown to be beyond what would otherwise be the indemnifying party's legal obligation then the consequences can be very serious, e.g. picking up the negligence of the owner.

The common law of England (i.e. the ancient unwritten law of the kingdom) is based on judicial decisions and it is very important in drafting clauses for a legal document for the drafter to be aware of how 'words' are currently interpreted by the courts. Statute law, on the other hand, equates with an Act of Parliament or regulation made thereunder and can be so drafted that it can override or otherwise the common law of the country.

8.10.1 Unfair Contract Terms Act, 1977

The passing of the Unfair Contract Terms Act 1977 and its subsequent enactment, was important in denying a person the right to exclude or restrict his or her liability by the use of contract clauses for death or personal injury resulting from negligence. The strongest protection is given by the Act to persons who deal as 'consumers', though those dealing otherwise than as consumers, e.g. where goods are brought for use in a business, are covered.

The Act also relates to loss or damage allowing the right to contest the terms of a clause, which is felt to be unfair or unreasonable. However, care has to be taken to ensure the Act can be applied to the particular contract.

The Act also deals with 'written standard terms'. Success in practice in contesting a contract term may well depend on a legal argument demonstrating the power of one party to impose what are deemed to be unfair and unreasonable terms on the other party as a result, for example, of having a monopoly position.

It is interesting to note that in an attempt to treat the ICE conditions in this way the court held that the general conditions were not a partisan document. However, it would be unwise of any contractor undertaking a construction project to enter into a contract embracing extremely onerous indemnity clauses on the basis of relying on the Unfair Contract Terms Act.

8.11 PROBLEM AREAS

Past case law has shown that difficulty can arise in the interpretation of a contract where, for example, the indemnity clause relating to property is not specific in its definition of the property to which it is intended to relate.

The contract may well have a specific clause relating to the contractor's obligations to carry out and complete the works; this clause does not limit that responsibility, unlike the indemnity clause, to the negligence of the contractor. It is in consequence very important to ensure when studying the clauses of the contract that a possible conflict of this nature is avoided.

Another point that can often arise in an indemnity clause is that its wording may not allow for the situation of contributory negligence, i.e. the negligent acts of other parties. For example, a clause wording may refer to 'unless due to any act or neglect of the contractor' and so call for the right to an indemnity where otherwise there may be a right to contribution. Whereas if those words read 'and to the extent the same is due to any act or neglect' the right to contribution would exist.

8.12 CURRENT PRACTICES

To a large extent the indemnity clauses contained within the standard forms of contract issued in the UK have been amended from time to time to eliminate ambiguities that previously existed and to recognize that one party may not be solely responsible for the event that has resulted in the indemnity being sought.

Nevertheless, it is common where the owner calls upon the contractor to provide an indemnity not to confine that indemnity to the negligence of the contractor but to embrace persons for whom the contractor is responsible, e.g. employees and subcontractors.

Without qualification it is likely that such terminology would confine the widening of the indemnity to the acts of persons for whom the contractor is responsible under the law of tort. So to counter such limitation it may be found the clause is worded to ensure there is a full description of such persons as, for example, in the JCT standard form of building contract 1980, clause 20.2.

8.12.1 The right to an indemnity

In general it should also be recognized that the obligation to indemnify does not normally arise until the loss against which the indemnity is claimed has actually been established. For example, the indemnifying party has been found responsible in law and damages (i.e. the amount of compensation) have been assessed and awarded. The timing of the claim being made could also raise the question of whether the seeking of the indemnity is made within the time limit allowed by law. For example, the law may require a summons to be issued and delivered to the defendant within a specific period for the claim to be allowed to be heard by the court.

8.12.2 Guilty until proved innocent

It is not uncommon within indemnity clauses to observe a wording that introduces 'a reversal of proof', i.e. the contractor is required to indemnify unless the contractor can show the injury loss or damage is the result of an act of the indemnified party. The argument for such a reversal of proof is that the contractor has custody of the site of the work and the owner, apart from acting through his engineer, has little presence.

Such reversal of proof does not form part of the law of tort but is imposed solely by contract. If called upon to accept only in relation to death or personal injury the position is not so serious for the reason explained above. However, a contractor should be very wary when such a clause extends to loss or damage to property and does not confine the indemnity to negligence.

An example of the exposure that could follow relates to the legal duty of one person to support the land of his neighbour. The duty is not confined to negligence, being more in the nature of a legal nuisance and under such a wide indemnity clause the contractor may well be required, where for instance there is subsidence on an adjoining property, to grant an indemnity to the owner in circumstances which would not but for the wideness of the clause call on the contractor to do so.

This is but one example which demonstrates the need for sound legal advice when entering into construction contracts for in this particular example it is not sufficient for the defendant to state that he took all reasonable care not to cause collapse of his neighbour's land, and in consequence the very wide indemnity granted by the contractor may be called upon.

8.13 AMENDED DOCUMENTS

It has already been mentioned that companies can prepare their own particular conditions of contract. In many instances the skeleton of the contract conditions can be traced back to one or more of the standard forms.

Nevertheless it is advisable to be extremely cautious when being presented with a 'typed' as opposed to a 'printed' wording on which to tender where that wording appears on face value to be identical to a standard form. It is most sensible to query why the owner has taken the trouble to so 'type'. Such an approach can well mean that minor but significant alterations have been made to what would otherwise be standard clauses. An example from the past is the omission from clause 20 of ICE conditions of contract 6th edition, of clause 20(2)(b), i.e. the excepted risk relating to design.

It has not been the intention, in referring to death or injury to persons, to specifically differentiate between employees and other persons. The indemnity clauses in a number of standard contracts will not do so although an example of separate clauses being introduced is shown in MF/1 clauses 43.4, 5 and 6.

Where indemnity clauses do not define a financial limit it means the indemnity is unlimited in amount.

8.14 NON-STANDARD CONDITIONS OF CONTRACT

In some standard forms reference is made to the limitation of liability, clause 44 of form MF/1 again being an example. Equally it may be found in tailor-made contracts for work offshore, such as offshore platforms, that the liability of the contractor to the owner has a financial limitation partly because there are few contractors capable of financing the size of damages that could be contemplated or the consequential loss that an owner could suffer or attempt to redress by way of liquidated damages. It is essential for the contractor to ensure that it is the contractor's liability which is limited and not just the amount for which the contractor is required to insure.

8.15 SOUND LEGAL ADVICE

It is not possible in this brief guidance to indemnity clauses to set out the entire legal position but if any party has any doubt as to the extent of the exposure, clearly such indemnity clauses should be referred to for legal advice.

It should be mentioned that indemnity clauses normally are unlimited in financial terms, unlike insurance policies where limits always apply.

It is hoped this introduction to indemnity clauses will at least have stressed the need to read carefully and to seek advice when necessary.

It is equally important that a contractor identifies insuring clauses whether they relate to the work being undertaken or to accidents that can occur during the construction period. Such clauses must be related to the indemnities being granted and again a need to ensure there is adequate internal or external specialist advice to ensure performance.

8.16 INSURING CLAUSES

Insuring clauses, as mentioned earlier, normally form part of construction contracts, but it would be very misleading to state that such clauses are drafted in a manner whereby the owner and the contractor can assume all incidents involving injury, loss or damage will be financed by the compensation granted by the insurer. For example the insured may have to bear a substantial part of any loss (e.g. 'excess' or 'deductible') or the terms of the policy may exclude certain types of losses.

Insurance clauses tend to relate to two forms of cover:

1. to indemnify the parties to the contract for the cost of repairing or reinstating work carried out, temporary works, materials and goods for incorporation therein and occasionally extended to constructional plant, equipment and site accommodation;
2. to ensure the contractor has liability insurance sufficient in scope and amount to support the indemnity the owner seeks from the contractor.

Because surety guarantees such as performance and retention bonds are issued by insurance companies as well as banks it is possible that such guarantees can be confused with the owner's insurance requirements. Bonds of this nature, however, are not insurance contracts.

The owner may seek from the contractor evidence that he has insurance to indemnify the contractor against defective design or defective designing depending upon the extent of the contractor's design liability. Both defective design and defective designing have been mentioned previously, the former extending to design failure that may be a breach of duty not normally placed on the contractor, e.g. unforeseen within the normal design duty of skill and care. As with indemnity clauses great care must be taken to understand clearly the obligations under insuring clauses and to identify if those obligations can be fulfilled.

An example of a 'failure to fulfil', i.e. to provide cover which is identical to the indemnity to be granted, would be where the indemnity clause is sufficiently wide as to embrace events that cause environmental pollution and where there is a corresponding insuring clause to arrange insurance to cover the indemnity. Clauses 22 (1) and (2) of ICE conditions of contract (6th edition) are an example. As it is now common for insurers to exclude from their public liability policies, injury, loss or damage due to gradual pollution it may not be possible to comply with the insuring clause and therefore as insurance cannot be provided it can be argued that the contractor is in breach of contract.

In the USA liability for pollution, such as the cost of clearing polluted sites, is a major problem. The influence of such problems has spread to Europe and as UK insurers are unable to secure adequate reinsurance, cover is now confined to what is termed 'sudden and accidental pollution', e.g. to secure an indemnity the discharge, dispersal, release, seepage or escape must arise from an identifiable single, sudden, unintended and unexpected event. Indeed at the time of preparing this chapter there was talk of excluding the risk of pollution completely from public liability policies.

An owner in purchasing a site for development would be well advised to embrace within his enquiries the possibility of pollution, and equally any contractor tendering must recognize the possible exposure especially if such a risk comes within the indemnity clause.

It is not uncommon to see the well-known non-insurable risks as exclusions in a policy of insurance, examples being war, sonic bangs and ionizing radiation from nuclear establishments or weapons. This position may be recognized in a construction contract by being risks retained by the owner or commonly known as excepted risks. The excepted risks specified in the contract should apply both to the indemnity and insuring clauses.

It must be appreciated that all insurance policies embrace terms of contract, which means for example there can be exclusions to the cover provided, conditions which are precedent to liability or indeed warranties. Details of conditions and warranties will be referred to later in this chapter.

It would be quite impossible to embrace within an insuring clause extensive detail of the cover required and which is known to be available. Therefore in general and as described in section 8.5 above there tends to be an acceptance by an owner that the cover granted adequately supports the indemnity clause(s).

Insurance cover is provided in the UK by a market made up of many hundreds of companies, mutuals and also Lloyd's.

Since the demise of the Fire Offices Committee there has been little standardization and therefore a description of a 'standard policy' is almost

impossible, each insurance company or Lloyd's underwriter issuing wordings that suit the company or underwriting syndicate. Most insurance
companies and underwriters will also be prepared to issue wordings tailor
made for a particular insured or contract.

8.17 CONDITIONS AND WARRANTIES

It is important both to recognize and understand a clause in an insurance
policy that is prefaced by such words as 'It is hereby warranted' and to distinguish from a 'Condition that is precedent to liability'. The difference is
normally explained by saying that a 'condition' requires an insurer to
demonstrate the breach influenced the loss, whereas in the case of a 'warranty' it is sufficient for the insurer to demonstrate the breach existed at the
time of the loss whether or not influencing the event, i.e. the incident that
has caused the claim.

An example would be a policy covering all risks of loss or damage to
property but containing a 'warranty' that when the premises are unoccu-
pied a burglar alarm must be in operation. If during unoccupancy the
premises catch fire and it was discovered the insured had failed to set the
alarm insurers would have no obligation to indemnify despite the alarm
having no relevance to the fire.

The point about conditions and warranties is most important in industrial engineering work as often there are bound to be hot workings taking
place. A vivid example was where the contractor sought a corresponding
indemnity to the one he had given in the main contract but failed to check
that the subcontractor's public liability policy embraced a 'heat treat warranty'. That warranty called for specific precautions to be taken but after a
serious fire had occurred damaging the existing plant the subcontractor was
found to be in breach and his insurer refused to grant an indemnity. This
left the contractor in the unfortunate position of having to fulfil the indemnity the contractor had given to the owner which indemnity embraced the
acts of subcontractors. It need hardly be stated that the subcontractor without recourse to insurance was not able to compensate the contractor for the
very large sum involved.

The obligation to insure the work being undertaken and possibly the
existing plant that is being worked upon can be either that of the contractor
or of the owner or indeed divided. It is to be hoped that the insurance being
arranged protects at least both the main parties to the contract. The scope of
that insurance may be clearly defined by specifying the perils against
which insurance is required. Problems can arise when reference is made to
all risks insurance and there is no clear definition in the contract of any
exceptions to the cover that will be acceptable.

An interesting point at law arises when in a subcontract the contractor
extends the benefit of the insurance he has arranged under the main contract to the subcontractor, which at the time of an occurrence is discovered
to be a unilateral decision in that the owner has not so approved the extension of the cover within the main contract. Hence the need to ensure the
main and subcontracts are complementary.

Very often either within an insuring clause or by way of an appendix to the form of tender the owner specifies the limit of indemnity to be granted under the public liability policy. It is not uncommon to find a totally unfounded belief that the nomination of such an amount equates with a limitation of liability. Where the sum specified by the owner is low then the contractor must ensure adequate protection and hence the contractor should deem the figure a minimum and select a limit of liability that is nearer the exposure faced.

Occasionally an owner will provide the public liability insurance and extend the benefit of the insurance to the contractor. This is likely to occur with a very large project and such insurance may be combined with material loss cover. Again, where the contractor has to place reliance on such insurance, where normally the liability for the work and indemnity would remain with the contractor, then great care should be taken to ensure at tender stage that the cover being provided is adequate.

8.17.1 Subrogation

Subrogation, which is a common law provision, can be defined as the right which one person has of standing in the place of another and of availing himself of the rights and remedies of that other person. The reason for mentioning this doctrine is that as most insurance policies contain a 'subrogation' clause there is a tendency to introduce the doctrine into construction contracts.

However, before developing this point it should be mentioned that in common law an insurer cannot compel the insured to allow him to exercise subrogation rights against a third party until the insurer has first paid the claim, unless the policy provides otherwise.

The introduction of a subrogation clause and its specific wording does prevent the insurer's claim against the third party being prejudiced by delay. The subrogated claim is incidentally taken in the name of the insured, it being part of the insured's duty to assist the insurer in the conduct of the claim.

8.17.2 Rights of recourse

Clearly the right to subrogation cannot apply against an insured for if an insurer undertakes to indemnify the insured then the insurer cannot thereafter set out to recover the compensation on the grounds, for example, that the insured had been negligent. In other words, an insurer having paid a claim cannot take legal action against its insured to recover the money the insurer has paid out.

8.17.3 Introduction into contract forms

The JCT, because of past case law, amended the definition of 'joint names policy' in clauses 1.3 and 22 of the standard form to read:

> Joint Names Policy: a policy of insurance which includes the Employer and the contractor as the insured *and under which insurers have no*

right of recourse against any person named as an insured, or pursuant to clause 22.3 recognized as an insured thereunder.

The words in italics (written as such in this chapter solely to highlight a point) are unnecessary and are really introduced for the comfort of parties not understanding the law of subrogation which has been explained above. It should be mentioned that the words 'which includes' are deliberate in that there may be other insured parties.

8.17.4 Other uses

Again referring to the JCT standard form the reference to subrogation is utilized in clause 22, as providing an option to grant to subcontractors certain benefits of the insurances arranged under clause 22, whereby the insurers can either add the subcontractors as insured parties or 'waive their rights of subrogation'.

8.17.5 Principal's clause

In a similar vein to the waiving of subrogation rights reference may be introduced into the insuring clauses of construction contracts and in particular to clauses dealing with liability cover to the effect that the policy should embrace a 'principal's clause'. This really equates with adding the name of any principal, e.g. the owner, with whom the insured is in contract and granting to that principal an indemnity.

It is important to understand that this does not necessarily provide the principal with the full rights of a 'joint insured'. However, the policy can be drafted to so provide.

Again it is possible to turn to the JCT standard form to indicate the limitation for under clause 21.1.1.2. the words read:

> shall indemnify the Employer [i.e. the principal] in like manner to the contractor but only to the extent that the contractor may be liable to indemnify the Employer under this Contract.

8.18 TYPES OF INSURANCE

8.18.1 Employers' liability (EL)

In simple terms this form of insurance sets out to indemnify an owner (i.e. employer) against a claim by an employee for death, injury or illness that arises out of and in the course of that person's employment.

A typical example of a requirement for the contractor to provide EL insurance can be found in MF/1 clause 47.5, but ICE (6th edition) is rather more interesting. The main feature of an EL policy is that the indemnity relates to the words 'arising out of **and** in the course of'. ICE attempts to distinguish between third party insurance and the indemnity in relation to employees. Unfortunately in the 6th edition the words in clause 23 read '(other than any operative or other person in the employment of the contractor)' so excluding from the obligation to insure 'arising out of **or** in the course of' circumstances which form part of the public liability cover.

Mention should be made of the difference in the way wording is drafted under an EL and a PL policy. In an EL policy the injury or illness must be 'caused' during the period of insurance and in the case of a PL policy the injury or damage must 'occur' during the period of insurance. In this way the EL policy provides cover even if the illness, e.g. an industrial disease, does not manifest itself immediately.

ICE conditions limit the reference to employees in clause 24 to an indemnity and do not require insurance. Doubtless this is due to the fact that such insurance in the UK is compulsory.

Since regulations were passed under the Employers' Liability (Compulsory Insurance) Act 1969 (which itself may be amended), owners, i.e. employers (with certain exceptions), have been required to insure their legal liability for injury to employees which arise out of and in the course of their employment. Hence the possible absence, as has been mentioned, of a construction contract containing an obligation to insure.

This demonstrates how varied are the insuring conditions in construction contracts. Complications may arise in that the definition of 'employee' in any EL policy can now extend beyond a person under a contract of apprenticeship or service, which means ensuring there is no gap between the cover granted under the EL policy and the public liability policy. Examples would be:

- a person hired or borrowed from another employer;
- a labour master or a person supplied by him or any person supplied by a labour only subcontractor;
- a self-employed person;
- a person supplied by a local authority, e.g. under a work experience training programme or a youth training scheme.

8.18.2 Value of the cover to be provided

An EL policy has had for very many years an unlimited indemnity, whereas a public liability policy will have a limit selected by the contractor or otherwise increased to conform with the limit stated in the construction contract. From the beginning of 1995 the insurance market reduced the limit of indemnity under the EL policy to a figure of £10 000 000, in respect of any one occurrence (and possibly for a series of claims against the insured arising out of any one occurrence). 'Occurrence' may in some policies be amended to read 'cause' or 'event'.

The decision is partly due to the *Piper Alpha* experience, which highlighted the catastrophic exposure of some risks, and the subsequent lack of ability by insurance companies either to buy at all or purchase reinsurance at a reasonable price.

The Employers' (Compulsory Insurance) Act 1969 will be of no assistance, for the limit specified within that Act is £2000 000 and despite considerable representation the government (at the time of going to print) has not disclosed any intention of increasing that value, the value having been proposed over 20 years ago.

In fact where, say, a public limited company purchased EL insurance under one policy but embracing the parent and all subsidiaries as the insured, the limit to £10 000 000 any one occurrence could have been a breach of the Act in that all employers are required to insure for a minimum of £2000 000.

However, by the Employers' Liability (Compulsory Insurance) General (Amendment) Regulation 1994 the government has avoided this breach, in that a 'group' policy with a limit of not less than £2000 000 complies with the Act.

Where the contractor is required by the terms of contract to procure insurance against injury to persons in general, and such terms nominate a limit of indemnity exceeding £10 000 000, it is possible either through the primary insurer or through an insurance market known as 'an excess of loss' market to purchase the balance of the cover required, i.e. the difference in the limit of indemnity required and the £10 000 000 granted under the primary EL policy.

It is important to point out that, again due to the *Piper Alpha* disaster, the EL limit of indemnity in relation to offshore work has been for sometime limited possibly to as little as £2000 000, with difficulty in securing 'excess of loss' cover. Otherwise with the use of the excess of loss market limits up to £100 000 000 for any one occurrence are available if such an amount is stipulated by contract or otherwise felt a necessary by the contractor.

8.18.3 Public liability (PL)

Briefly this form of insurance provides an indemnity against claims for the accidental death of or bodily injury to persons or the accidental loss of or damage to property occurring during the period of insurance and arising out of the contractor's business (as defined). The indemnity in many cases is now extended to take into account other forms of possible legal liability such as interference with traffic, obstruction, loss of amenities, trespass or nuisance.

In clause 18.1.1 (third paragraph) the variation in the wording of the 'operative clause', i.e. the clause embracing the description of the indemnity to be granted under EL and PL policies was explained. It should be added that in addition to the loss under a PL policy having to 'occur' during the period of insurance the injury or damage must be of an 'accidental' nature. In other words insurers do not intend to indemnify an insured who intentionally causes injury or damage.

Again there is no standard industry form of PL policy and although for manufacturing companies public and products liability may be dealt with separately, in the case of a contractor where the supply of materials and goods is simply for incorporation in the work it is preferable to contain the two forms of cover in one common wording. For the purpose of this chapter PL will be deemed to embrace products liability.

Examples of contract clauses specifying public liability or third party insurance (the two terms being synonymous) are MF/1 clause 47.4, ICE (6th edition) clause 23 and the Institution of Chemical Engineers Red Book clause 32. A study of those clauses will give the reader some idea of the different way in which such clauses are drafted.

Normally the term 'an indemnity against legal liability' in a policy of insurance will cover both liabilities in tort and statutory exposures. When in referring to indemnity clauses it has been made clear that a contractor can incur contractual liability, it is important that the PL policy is adequately worded to address these additional liabilities.

Although the construction contract requires the contractor to obtain insurance in respect of the indemnity he is required to give to the owner, in making his insurance arrangements the instruction by the contractor to his broker/insurer will not be so limited.

Apart from the indemnity there is a duty owed beyond the owner to other parties and the contractor should insure his liability for injury to other parties or loss or damage to their property.

In section 8.5 it was explained that an owner may choose to arrange insurance to protect both the owner and contractor. In such circumstances it will generally be found the cover provided will not be confined to any indemnity granted but, as mentioned above, will include protection against the contractor's legal liability to other parties. It would necessitate a chapter in itself and of a very extensive nature in order to provide guidance on suitable wording.

The wording, however, should take into account that:

- the 'operative' clause is carefully drafted;
- the cover includes products liability;
- contractual liability is not excluded;
- if there is any reference in the exclusions to design or to the cost of rectifying defective materials or workmanship such exclusions must be fully understood;
- any exclusion relating to property in the contractor's custody or control is qualified by reference to property being worked upon and also that occupied for the purpose of the work;
- where by contract the owner seeks protection under the PL policy this is appropriately arranged by way of a joint insured or a 'principal's' clause, e.g. by adding the name of the owner as an insured party or by introducing a clause granting an indemnity to the contractor's principals, one being in this case the owner; and
- as required by, e.g. ICE (6th edition) clause 23(2) a suitable 'cross liability' clause is embraced.

This is only a short reference to some of the aspects of a PL policy and must not be taken as a comprehensive guide but simply be utilized to recognize the need for advice from a specialist who understands liability insurance very thoroughly.

Reference to PL policies cannot be completed without contemplating that the construction contract may include design of the plant. The terms of the contract may require the contractor to provide plant fit for its purpose or limit the design liability to that of a professional acting under a separate contract where the liability is intended to be restricted to 'reasonable skill and care'. It is most unlikely that the PL policy will provide adequate protection in either case and professional indemnity insurance will be needed.

8.18.4 Professional indemnity (PI)

If the contract seeks insurance to cover the design liability it is most unlikely that the PL policy together with the policy covering material loss will fulfil the requirement and there will be a need to provide 'professional indemnity' (PI) cover.

The form of policy (as will be explained later) differs from that of the 'pure' professional, e.g. consulting engineer and is generally known as a 'design and construct' PI policy.

Again it is not possible in this guide to refer to all the salient points to be considered when buying this form of cover short of saying that it is a very small and specialized market and a limited number of insurance brokers have the specialist knowledge required to arrange this cover.

To give one example a professional would expect to be indemnified against a claim that arises out of a negligent act, error or omission on his part. In the case of a failure occurring in the contractor's design before handover (i.e. before the construction work has been completed) there is no claim to be made by the owner and the contractor will simply be required to fulfil his contractual obligations as opposed to the owner needing to prove the failure is due to a negligent act, error or omission. In such circumstances the standard PI wording granting cover against legal liability arising out of negligent acts, errors or omissions would not so accommodate a requirement simply to fulfil the contract and the wording of the policy must be widened.

Equally the contractor may well subcontract part of the design so he must make sure the reference in the policy of insurance to 'a negligent act, error or omission' is not limited to those of the contractor but ensure the policy provides the contractor with an indemnity arising out of such failures of the designer to whom the contractor subcontracted the design.

Having explained that an EL policy is on a 'causation' basis and a PL policy on an 'occurrence' basis , variation is again seen in a PI policy in that it is underwritten on a 'claims made' basis, i.e. the claim against the insured must be made during the period of insurance.

PL and PI policies have numerous exclusions, far too many to refer to in detail and hence the need for specialist advice.

Most PL policies have an exclusion relating to 'penalties and liquidated damages': these forms of liability will be referred to later in this chapter.

In Chapter 7 reference is made to the Housing Grants, Construction and Regeneration Act 1996 and in particular Part II Construction Contracts. The Act is so newly on the statute books that there is no practical experience of its operation but certainly PI underwriters are very concerned that if an adjudicator's decision made within 28 days of the dispute being referred is considered final (unless brought to arbitration but only at the end of the contract) they may be required to indemnify their insured immediately. PI premiums at the moment certainly take into account the lengthy period during which the majority of such premiums will be invested before any court action is finally resolved.

8.18.5 Subcontractors' insurance

Before discussing insurance of the work undertaken, completion of the reference to EL and PL insurance must culminate in the mention of subcontractors' insurances.

There is no standard format in construction contracts and it may be argued by an owner that there need be no reference to subcontractors' insurance in that the indemnity given by the contractor embraces the acts of subcontractors and the contractor is required to insure against such liabilities.

Other forms may well recite in a similar manner but equally embrace such words as 'the Contractor's obligation to insure shall be satisfied if the subcontractor shall have insured against the liability in respect of ...'

Many subcontractors will have in existence, possibly on an annual basis, both EL and PL policies and will doubtless have costed such insurance in producing a subcontract tender price. For the contractor to extend his policies to provide *all* subcontractors with cover will undoubtedly be a duplication and one for which the owner is likely to pay.

Nevertheless, it is very important that the insurance arranged by the contractor covers liability for the acts of all the subcontractors whether labour only or otherwise. In law the contractor, in certain circumstances, can incur liability for the acts of his subcontractors, an example being work of a dangerous nature.

However, where reference is made to subcontractors in the main contract it is incumbent on the contractor to ensure the subcontract conditions adopted are quite explicit in the insurance the subcontractor is required to provide and to ensure that cover is effected.

To protect the contractor's own claims experience and minimize future premiums it is important to check that a subcontractor is adequately insured. If there is a failure and the owner requires the contractor to provide compensation by way of the indemnity clause forming part of the main contract, then such an award if reimbursed by insurers will prejudice the contractor's claims experience.

An interesting situation is when the main contract may specify a very large PL limit of indemnity, e.g. £25 000 000, and one that the contractor is required to ensure subcontractors also provide. Many subcontractors would find it hard to procure such a limit and in so doing it may be found that the aggregate of the prices such subcontractors are charged for cover in excess of their normal limit will prove prohibitive. It may be preferable financially for the contractor to arrange insurance covering all subcontractors in excess of a lower limit that is specified in subcontracts. Such insurance would need internal specialist skill or the assistance of an insurance broker dealing in this form of cover.

Reference to a contractor's liability for the subcontract works and to insure them against loss or damage will be dealt with later in this chapter.

A subcontractor's design liability for the permanent works is a subject that causes great concern. Normally a contractor has no desire to embrace within a construction only contract that is limited to workmanship and materials, design by reason that a specialist subcontractor has designed an element of the works. If such a situation arises the contractor must view whether he can impose upon the specialist subcontractor a liability greater than the normal one of 'skill and care' and seek one of fitness for purpose. Fitness for purpose can be very difficult to impose when a specialist contractor designs only part of the whole. Equally the contractor will need to decide if the design liability must be supported by what is commonly known as a 'professional indemnity' (PI) policy.

PI cover in relation to the contractor has already been discussed and in passing down the responsibility to a subcontractor similar considerations arise. Where the contractor is not designing the works it may be possible to attempt to ensure the specialist subcontractor's design responsibilities are specified in a separate contract directly with the owner.

8.18.6 Material loss

Two fundamental questions arise, namely which party carries the risk of loss or damage to the works and which party is required to insure against material loss or damage. The risk of workmanship and materials is almost certain to rest with the contractor and also design responsibility where so required by the form of contract. Exceptions are likely to be embraced and retained by the owner — examples being war, riot and civil commotion.

Whichever party is required to insure it must be remembered that normally the cover will set out to protect both the owner's and contractor's interest in the works. Whether subcontractors are also protected varies greatly and depends on the form of contract utilized.

The wording of insuring clauses varies greatly betweeen the forms of construction contracts; some clauses require very wide forms of cover while others are more restricted. Equally, some clauses are clear and others ambiguous. However, it should be borne in mind that the obligations placed on a contractor to perform in many cases will extend beyond the forms of loss, damage or liability insurance which are required to be arranged under the contract. On the other hand the wording of the insuring clause may be so loose as to be ambiguous. For example form MF/1 in clause 13.1 refers to: 'with due care and diligence, design, manufacture, deliver on Site, erect and test the Plant, execute the Works and carry out the Tests on Completion'. This example embraces very onerous liabilities and although there may be excepted risks (i.e. risks retained by the owner) it is necessary to compare the responsibility with the obligations to insure.

In form MF/1 clause 47.1 states 'The Contractor shall in the joint names of the contractor and the Purchaser insure the Works and contractor's equipment ... for their full replacement value against all loss or damage from whatever cause arising, other than the Purchaser's Risks'.

Clause 45.1 refers to the 'Purchaser's Risks' which in précis form relate to the purchaser's or his engineer's design, use and occupation of the site by the works, etc., rights of way and light and so on, damage which is inevitable excluding the contractor's method of construction, use of the works by the purchaser, acts of the engineer or the purchaser, etc., and *force majeure* except to the extent insured by reason of clause 47.

It is not difficult to identify an element of ambiguity by the qualification to *force majeure* for whether some of the risks specified in clause 45.1 are insurable will depend upon the insurer so selected.

Equally the reference in clause 47.1 to 'whatever cause arising' does seem to embrace all loss or damage due to design for which the contractor is responsible, such a broad definition of cover being unlikely to be achieved within a material loss policy.

The example provided is not intended to be a criticism of a specific form of contract wording but to demonstrate the need to fully understand insuring clauses, to identify any ambiguities and if necessary secure clarification. Equally, where the contractor's responsibility is wider than the insurance required under the contract, it is necessary to ensure, if possible, the cover is also widened.

It would be impossible in this chapter to provide a detailed recitation of material loss insurance. However, all risks cover provided for industrial engineering works as opposed to building and civil engineering works is normally known as engineering all risks (EAR). For building and civil engineering works the term utilized is contractors' all risks (CAR). A reader is bound to question the position when the project embraces both forms of work. The unofficial and general guide in the Insurance market is to identify the percentage of work that is industrial engineering and if this exceeds 50% of the total value treat as EAR but otherwise as CAR.

Each form of policy sets out to provide an indemnity against 'damage' to specified property and in some cases 'damage' will be defined as 'loss or damage' and in others 'physical loss, destruction or damage'.

Underwriters are prepared to embrace within the cover a selection of property according to the insured's wishes, examples being the permanent works, temporary works, constructional plant and equipment either owned or hired, site huts or other structures such as caravans, tools and personal effects of employees. That property as well as being covered on site can be extended, in certain circumstances, to off-site storage, while at other premises and while in transit.

The breadth of cover afforded also varies in relation to the insurance provided during the period known as 'the defects liability period' or 'maintenance period'. The aim of such cover is to indemnify the contractor against damage occurring during such period but resulting from a happening prior to 'handover' and also to damage caused while undertaking work during that same period. The wording of such an extension in relation to industrial engineering projects can vary being in some cases restricted to a defect occurring during the construction period but sometimes even when the defect arose during manufacturing.

One further warning relates to a construction contract that may call upon the contractor to continue full CAR or EAR cover for a period after 'handover'. An example is in MF/1 clause 47.1 seeking that extension for a period of 14 days. Certainly it is to be preferred that the owner arranges insurance of the works after they are handed over.

Although accommodating, insurers underwriting all risks insurance for construction risks are certainly not happy to entertain the risk after handover when there is permanent use of part or whole of the project, especially as the rating structure is not designed for such an exposure, e.g. the rate is calculated as applying to a building or plant in course of construction and not, for instance, as a plant processing highly flammable liquids or a building in which highly combustible materials are being stored.

It is important for a contractor to recognize the excess or 'deductible' imposed by an insurer when granting all risks cover. The sum may well be reasonable for the general type of loss such as a fire, possibly higher for storm or flood damage but in the case of loss due to defects in design, material or workmanship it can be a very substantial sum especially if the 'design' cover granted is wide. A figure of £50 000 or even a £100 000 can be contemplated.

To attempt to deal in detail in this chapter with 'offshore' works would be quite impossible, but where such work embraces ocean tows very great care has to be taken to ensure the marine cover and the construction cover, if written in separate policies, do not duplicate or leave gaps.

Insurers are normally willing to cover the interest of a number of parties (i.e. include various names as the insured under one policy) including the owner, contractor and subcontractors. It is often difficult to add the name of the engineer and other professionals appointed by the owner or that of manufacturers of plant for installation. The reason is that insurers seek to retain rights of subrogation against them.

No insurer will simply issue a policy covering all risks of loss or damage of whatsoever nature to the insured property and hence with such a wide type of wording the policy will bear exclusions to the cover provided.

To ensure the scope of the cover is fully understood the contractor will need to utilize appropriate internal skills or engage the services of a specialist insurance broker.

The exclusions and qualification to 'all risks' forms of cover are very extensive and as there is no standard wording alternative wordings exist in the Insurance market. Various extensions to the normal type of cover are also available.

Some of the more common extensions are:

- professional fees;
- debris removal;
- escalation of value;
- automatic reinstatement after loss;

- expediting expenses;
- local authorities' requirements;
- immobilization without physical damage of, e.g. constructional plant;
- water borne craft;
- defective design;
- inventory losses;
- inevitable damage;
- cessation of work;
- breakdown during testing and commissioning.

Testing and commissioning cover in relation to industrial engineering projects is a very important aspect of the insurance and it will normally be found the periods for testing are limited and cover for commissioning depends upon whether the contractor is called upon to commission with feedstocks utilized or simply required to assist. Insurers' great concern is the provision of breakdown cover if the plant is not new, or even more importantly is of a prototype or experimental nature or indeed of unproven design. There is no standard wording relating to loss or damage due to design, the extent of the cover granted being reflected in the premium paid.

8.18.7 Terrorism

Where the form of contract seeks 'all risks' cover then those words embrace the risk of 'terrorism'. There is an agreed definition of the term terrorism such as an act of terrorism having to result in fire or explosion damage.

For insurers to provide cover for large sums insured they normally reinsure with the Pool Reinsurance Company with the government underwriting as 'reinsurer of last resort'. The terms of the reinsurance granted by the Pool Re. forbid an insurance company subscribing to the reinsurance to grant terrorism cover for a period exceeding 12 months. Under a construction contract the requirement to insure the works is normally for the full construction period. Hence where a construction contract exceeds a 12-month period there is clearly a breach if the contractor does not provide the full material loss insurance for the whole period of the contract which will not be possible as the cover against terrorism will be for a period less than the full construction period.

Certainly the JCT have made provision for this problem by amendment of their standard forms of contract; but it is necessary to be alerted to this problem if utilizing other forms of contract.

The Pool Re. announced on the 13 June 1996 that the review of premium rates undertaken following the ceasefire by the IRA and others was to be discontinued now that it had been established that the estimated losses following the South Quay explosion on the 9 February 1996 would exceed £75 million, being the 'cut off' point at which an insured would be called upon to pay the 40% deferred premium. The bombing in the Centre of Manchester immediately after this statement was made by the Pool Re. makes it clear that the losses the Pool Re. will now be called upon to pay will greatly exceed the 'cut off' point of £75 million.

Therefore the concession granted by the Pool Re. applying to EAR and CAR policies, namely that where only 60% of the premium has been paid for a policy of short duration and that policy has expired by the time the Pool Re. recognized losses have actually exceeded the 'cut off' figure then the insured under such a policy would not be required to pay the balance of the premium, i.e. the 40% will no longer apply to new policies.

8.18.8 Variations to standard forms

Variations to the standard practice of requiring the contractor to insure the industrial works can also be experienced. For instance, the project may be of such dimensions that the owner will decide to provide the material loss insurance. The need then is for the contractor to establish the true extent of cover at the time of submitting a tender for undoubtedly the contractor will not be relieved of his responsibility for the works and in any case may be called upon to meet claims falling within the excess (i.e. the first amount of any claim) under the policy.

Equally a form of contract known as project or construction management may be utilized, with separate sections of work being undertaken by various main (i.e. trade) contractors. In such a case the project manager/owner may arrange the material loss insurance or seek insurance in accordance with each individual trade contract. Where the project manager/owner provides the insurance on the works it may well continue until final completion of the project; but a 'trade' contractor should be alerted to the fact the benefit of the cover afforded under the individual 'trade' contract may well cease at the completion of that individual contract. If damage occurs after the trade work has been handed over the trade contractor cannot rely on the project insurance for an indemnity, save for any extension providing cover during any defects liability period.

Project or construction management contracts are common in the oil and gas industries. The contractor is most likely to be made responsible for the work being undertaken irrespective of which party insures. However, in relation to constructional plant and equipment utilized by the contractor for carrying out the works, there may be an obligation to accept full responsibility for any loss or damage even without any right to seek recourse from the owner if the owner is found to be responsible for such loss or damage. The form of contract utilized will state any requirement for such constructional plant and equipment to be insured.

In considering material loss insurance consideration must be given to existing property and its contents when the project is to be carried out in or on existing plant, i.e. who is responsible for it and who insures it. It is essential that the main contract is read with this particular exposure in mind not only in relation to the material loss but (as the insurance market as opposed to English law describes it) the consequential loss that can flow from such material damage. It is particularly important to compare any clause requiring insurance with the corresponding indemnity clause. If there is ambiguity argument may pursue, the case for the contractor being the belief that any undertaking by the owner to insure embraced not only material loss of the existing plant and structures but also any consequential loss that arose. The argument by the owner would be to the contrary.

Working in some of the major UK chemical plants or oil refineries can introduce an exposure of such magnitude that it is beyond the reasonable financial ability of even some of the largest contractors and hence waivers of liability or financial limit to that liability, joint names insurance on such plant or installation and concessions in relation to consequential loss must be viewed at the time of tendering.

8.19 OTHER FORMS OF INSURANCE

8.19.1 Performance bonds

Both MF/1 and ICE forms of contract make provision for the employer, i.e. owner, to seek a performance bond from the contractor.

Although it has already been stated, performance bonds are not insurance policies and it is important for a contractor to ensure the terms of the bond or guarantee are fully understood especially that the wording does not equate with an 'on demand' bond. An on demand bond would normally allow the beneficiary (i.e. the owner) to call upon the surety to pay the amount of the bond or other appropriate sum with or without evidence of the contractor's failure to perform.

The courts have recently been critical of the forms of wording adopted partly because of their lack of clarity, and changes can well be envisaged. For example, someone may well ask why the contractor has to be jointly bound under the bond.

It is also interesting to study the attitude of the banks and insurance companies towards bonds. The banks tend to require 'financial certainty' so that once the call on the bond is shown to be in accordance with the wording the demand can be honoured. Insurance companies, on the other hand, wish to secure the full benefits of a 'surety' to the extent of undertaking performance if necessary. This distinction in approach mainly results from banks not wishing to undertake the full responsibilities and duties of a surety, arguing that such responsibilities and duties are not part of the skills they wish to possess or perform. Insurers, on the other hand, not normally being in a position immediately 'to debit an account' with the amount of any financial guarantee paid, prefer to have the right to undertake the full duties of a surety and if necessary complete the performance of the work rather than simply meeting a financial claim under the bond.

Care must equally be taken in studying the 'counter indemnity' required by the 'bondsman' either from the contractor or, if the contractor is a subsidiary company, possibly from the contractor's parent company.

Performance bonds, although the primary form of bond, are not the sole form and, for example, bonds can be granted in return for the release of what would otherwise have been retention moneys and also for advance payments.

8.19.2 Parent company guarantees

Mention has been made that performance bonds are not policies of insurance, but as it has been seen suitable to refer to such instruments and indeed to counter indemnities, mention shall also therefore be made of 'parent company guarantees'. Where a contractor has no parent company then such guarantees are inappropriate. Equally, parent companies are reticent about guaranteeing the performance of a subsidiary, partly because this is a financial exposure to the group of companies that is not welcomed.

Nevertheless there are circumstances where in order to secure the award of a contract a guarantee of this nature is given. That guarantee may or may not be in addition to a bond issued by a surety.

There is a tendency to utilize the term 'guarantee' loosely but in law a guarantee is given on behalf of another party, it not being possible for a contractor 'to guarantee itself'. A parent company guarantee can also be known as a group company guarantee.

8.19.3 Business interruption

Some insurers are willing to extend EAR or CAR policies to cover the consequential loss flowing directly from material damage for which an indemnity is granted. This could relate to additional payments the owner is required to make to the contractor or additional costs the contractor must bear without recompense from the owner.

Such consequential loss can relate to 'idle or standing time' which is not recoverable under the contract or to the additional cost of completing work which has not been started or having been started is delayed as a result of the insured damage.

Many contractors would not insure against business interruption but it is as well to know that if the individual or annual EAR or CAR policy does not provide this cover there is within the Insurance market a specialist section underwriting what is know as 'business interruption forms of insurance'. This specialist section will also accommodate the owner's interest by providing cover for loss of turnover or profit.

8.19.4 Credit insurance

It is not the intention in this chapter to deal with export credit guarantees, but it is possible for a contractor to secure insurance which is aimed at indemnifying in the event of the insolvency of the owner.

Credit insurance can be arranged in many forms; in relation to a specific construction contract there is a tendency to seek such cover if the contractor has doubts about the financial standing of the owner; but in such circumstances it will be found that the underwriter of such cover is reticent to grant insurance equally recognizing that financial instability.

8.19.5 Engineering services

The services supplied by the specialist engineering insurance companies to comply, for example, with statutory regulations in relation to the inspection and testing of pressure systems, lifting plant and tackle is not insurance and does not form part of this chapter. However, it is important to make reference to such services which can be coupled with insurance. Equally, these companies will undertake the inspection and testing of large vessels or other plant to be utilized in engineering projects during both design and manufacture and also to provide cover during that manufacture and transit. The services afforded are very extensive, including such services as destructive and non-destructive testing.

8.19.6 Annual covers

Although in the main this chapter has aimed at dealing with insurances required in relation to specific construction contracts, clearly a number of the covers mentioned are arranged by a contractor on an annual basis, for it would be uneconomic in time alone to attempt to place individual covers for a multitude of small and medium engineering contracts and subcontracts that could be secured in any period of 12 months.

These annual types of insurance will also deal with the property at permanent establishments, motor vehicles, computer equipment/net works and the 'hacking' thereof, dishonesty by staff, directors' and officers' indemnities, political risks and employee benefits, to quote but a few examples.

8.19.7 Ascertained or liquidated damages

English law, and in particular the law of tort, requires a claimant who has suffered a loss due to breach of duty owed by another party to prove his financial loss and as such that sum is likely to be awarded as damages.

However, in contract it is acceptable for both parties to the contract to accept in the event of certain circumstances occurring, that a sum which was agreed at the time of entering into the contract will be the compensation paid by one party to the other.

In the majority of contracts for building or civil engineering works the sum known either as ascertained or liquidated damages is payable when the contractor does not complete the work being undertaken in the time prescribed. The figure agreed can be a weekly or a monthly sum with or without a maximum limit.

It would be inappropriate to seek payment in all circumstances causing delay and the contract will specify certain circumstances where an extension of time is granted instead of the liquidated sum being paid (or deducted from moneys owing to the contractor). Examples would be *force majeure* or a major peril such as fire or explosion.

It is known for industrial engineering contracts to extend liquidated damages to cover lack of performance of the plant, inadequate product or output, excess use of raw materials or the consumption of power, stack emissions or possibly excess waste product or quality of that waste.

Both building and engineering contracts require the contractor to perform in accordance with the contract, but in relation to liquidated damages it could be argued that an engineering contract is more severe in that the specification may require very keen limits on product quality and quantity and at an expected production cost.

In general the law in relation to such agreed damages is complicated for even if the contractor has agreed a sum this does not remove the contractor's right to apply to the courts to have the sum recognized as a 'penalty' and thereby for payment to be overruled, although this would not deny the defendant then the right to seek unascertained damages (i.e. actual costs).

8.20 RISK MANAGEMENT

With the severity of loss made by the insurance market in the early part of the 1990s the need for a contractor to demonstrate the capability of its directors and staff to manage 'risk' has become more and more of a key factor not only as a sensible way of running a company but also in being able to secure reasonable insurance premiums. The board of any construction company should be well aware of this subject and indeed ensure risk management is fully practised.

It can be argued that risks have always been managed but this concept, now known as risk management, concentrates a board's attention on ensuring risks to which the company is exposed are identified, their elimination or reduction designed and controlled thereafter.

In particular, attention should be drawn to the Joint Code of Practice on the Protection from Fire of Construction Sites and Buildings Undergoing Renovations; insurers are likely to import the operation of this code into all risks construction policies and not necessarily confine it to buildings but also extend it to other forms of work. The code is normally only enforced where the original contract price exceeds a figure selected by the particular insurer as the minimum valued contract to which the underwriter will apply the code. Insurers tend not to apply the code to the smaller valued contract but there is no 'market' agreement to the minimum value which will be applied. This code in conjunction with the CDM regulations will certainly impose upon both designers and contractors very strict duties and obligations.

The CDM Regulations (The Construction [Design and Management] Regulations 1994) came into force on the 31 March 1995 and are possibly the most far reaching Health and Safety regulations influencing the construction industry that have ever been promulgated by Parliament.

8.21 CONSTRUCTING THE TEAM

It would be quite inappropriate to conclude this chapter without making reference to Sir Michael Latham's final report issued in July 1994 entitled *Joint Review of Procurement and Contractual Arrangements in the United Kingdom Construction Industry* for it lays great emphasis, among other things, on the value of standard forms of contract and the need for some organizations issuing standard forms to amalgamate. Indeed, a recommendation is made that the new engineering contract, with its core clauses, would be an appropriate form of contract for the construction industry to utilize. It should be made clear no attempt has been made in this chapter to refer in detail to this form of contract, which as indicated by a footnote in the report has been very little used to date.

However, a brief study of clause 8, 'Risk and insurance' in the second edition will not go amiss, for the wording is certainly likely to introduce legal argument and for insurers some difficulty in producing policies which comply. The basic concept of each party not only carrying risk but in the case of the contractor and subcontractors insuring that risk under separate policies (and the owner also doing so at his option) surely will result in greater complication than the forms of contract which seek to insure in joint names against specified risks and so avoid the need to identify where liability would otherwise rest.

Equally, core clause 6, 'Compensation events', which divides the owner's and contractor's risks, could again produce complication, especially when there is a serious fire and an allegation that the cause could rest with the contractor.

BIBLIOGRAPHY

Advanced Study Group 208A, *Construction and Erection Insurance*, The Insurance Institute of London, Aldermanbury, EC2B 7HY.

Anderson, Bickford Smith, Palmer, Redmond-Cooper, *Emsden's Construction Law*, Butterworth, looseleaf with continuous updating.

Bunni, Nigel G. (1986) *Construction Insurance*, Elsevier Science.

Duncan Wallace, I.N. (1995) *Hudson Building and Engineering Contracts*, 11th edn, Sweet and Maxwell.

Hickson, R.J. (1987) *Construction Insurance: Management and Claims*, E & F N Spon.

Levine, M. and Wood, J. (1991) *Construction Insurance and UK Construction Contracts*, Lloyd's of London Press.

Report of Study Teams on Auditors/Construction Industry/Insurance/Liability/Professionals/Surveyors (1989), *Professional Liability*, HMSO.

Wright, J.D. (1991) *Construction Insurance*, Witherby.

Contractor/ Subcontractor Selection

Mechanical installations at THORP nuclear processing facility Sellafield, England.

Facility Owner British Nuclear Fuels Ltd.

9.1 INTRODUCTION

The effective selection of appropriate contracting and subcontracting organizations to undertake the implementation phase of a project is vital to its success. This chapter seeks to explore the process for ensuring that the choice of best contractor is made in all the circumstances; however, since all projects are unique, the following should be regarded as a checklist and not a prescription for the selection of a contractor.

There are many factors to be considered in arriving at the decision which may be summarized as:

- capability;
- credibility;
- enthusiasm.

In addition to basic commercial and contractual considerations there are an increasing number of constraints to be observed in the selection process.

Private individuals and many privately owned organizations may operate with little outside interference, while many larger organizations, especially those with, or aspiring to, quality assurance accreditation will have corporate procurement procedures. It is recommended, however, that such procedures are based on the principles contained in the recognized codes of tendering practice, or the codes used directly, in order to give tenderers confidence in the fairness of the tendering process. Such codes include the NJCC Codes of Procedure for single stage selective tendering, two stage selective tendering and selective tendering for design and build.

Governmental organizations and utilities are subject to EC procurement directives which are intended to ensure fair and equitable tendering procedures, value for money and auditability. These directives are sufficiently influential to warrant the detailed summary which appears later in this chapter.

9.2 SELECTION STRATEGY

The main activities to be derived from a selection strategy are

- establishing selection criteria;
- pre-qualification;
- shortlisting;
- tender evaluation;
- negotiation/clarification;
- selection.

There can be no doubt that a selection process, based on an understanding of project objectives, carried out using a structured, analytical, objective approach and tempered using risk management techniques, can greatly assist in the choice of the most appropriate contractor.

9.3 ESTABLISHING SELECTION CRITERIA

Selection criteria should be established following an analysis of the principal aims of the project. These may include:

*   safety;
*   functionality;
*   programme;
*   customer satisfaction;
*   cost;
*   third party issues.

The above factors cannot all be of equal importance and establishing their relative merits and agreeing upon a hierarchy creates a framework for decision making throughout the project. The first set of decisions may well involve converting the aims into objectives, by setting units and means of measurement, parameters and time scales. These, in turn, are the foundation for value engineering and risk management.

Armed with objectives and alerted to probable risks and sensitivities, the team undertaking the evaluation can establish selection criteria against which to assess the suitability of potential contractors and subcontractors which may include:

*   relevant experience;
*   availability of resources;
*   in-house capability;
*   safety record;
*   quality assurance accreditation and systems;
*   current and projected workload;
*   financial stability.

9.4 PRE-QUALIFICATION

Pre-qualification is a selection process in which an extensive list of potential contractors/subcontractors is evaluated against certain criteria in order to establish a short list of those to be invited to submit a tender. Some organizations refer to this process as simply 'qualification', but the distinction is notional.

The purpose of the pre-qualification process is to select the contractors/subcontractors to be invited to tender. For certain works, e.g. work of a specialist nature, the need for pre-qualification may not arise as the number of companies capable of undertaking the work may be limited, in which case the market will determine the final tender list. However, this is not normally the case and pre-qualification, of some description, will have to take place.

The initial list may be drawn from:

*   expressions of interest submitted in response to published procurement notices including those required by the EC;

- registers of contractors/subcontractors held by the owner or another member of the procurement team. Although not universal within the industry the register concept provides an initial starting point at which to seek potential contractors/subcontractors. The details on any register will vary, but they will give some insight into a company and allow a quick initial appraisal. It may be that the register will be used as an 'initial selection' device or alternatively be used as a mailing list for obtaining expressions of interest;
- lists of organizations provided by trade organizations;
- direct or indirect knowledge of the organizations whether from other projects, word of mouth, articles in the trade press or advertisement. Even though the team may know a reasonable number of companies, it should consider searching for new companies which, perhaps due to the project locality, may be equally suitable.

The importance of pre-qualification as part of the process of selecting the contractor/subcontractor cannot be overemphasized. It is as a result of the pre-qualification process that the final tender list is drawn up and it is from this list that the successful contractor will be selected. It is therefore imperative that the final tender list should contain only those contractors/subcontractors that are experienced, capable and interested in tendering for the work.

The type of pre-qualification undertaken and its extent will depend on the project itself. It is unrealistic to expect the same amount of time and energy to be spent on pre-qualification for a £50 000 project compared with a £100 million project; however, the reason for pre-qualification remains the same.

The pre-qualification process for each contract takes into account the size, nature of the work, complexity, programme of work, location, likely form of contract, the owner, any statutory requirements, etc., and it should be carried out impartially and with the best interests of the project in mind. It is the responsibility of the project team to ensure that the final tender list and subsequently the successful contractor/subcontractor is not only technically, administratively and financially capable of undertaking the project but is also willing and enthusiastic.

9.4.1 Pre-qualification content

The pre-qualification will request general company information, name, address, contact persons, ultimate holding companies, etc., and will seek responses in order to satisfy the following typical pre-qualification criteria, derived from the general selection criteria covered above.

- Compliance: does the contractor comply with stated technical and professional requirements, e.g. inclusion on an approved list, membership of professional/trade organizations, etc.?
- Financial stability: is the contractor operating profitably? Are the financial burdens of a major project due to payment terms, retentions,

bonds, etc., likely to cause financial difficulties, under-resourcing of the work or even insolvency?

- Experience: what is the type of work normally undertaken? Does the contractor have experience of similar projects, particularly if the proposed project has specific challenges and requirements? Has the contractor worked for similar owners? What is the contractor's track record in the kind of works under consideration? Does the contractor have experience of similar prevailing climatic, geographical and cultural conditions? Does the contractor employ suitably experienced key personnel?
- Capacity/resources: what capacity is available considering current projects and outstanding tenders? What is the general availability of labour, plant and key personnel? Quality systems: can the contractor demonstrate a corporate commitment to the achievement of quality products?
- Health/safety: does the contractor have an acceptable health and safety policy and procedures including Construction (Design and Management) Regulations 1994, and an acceptable safety record?
- Industrial relations: does the contractor have an acceptable policy and record?
- Project specific: does the contractor have experience of matters which are peculiar to the work, e.g. special construction techniques or design knowledge?

The relative importance attached to each of the various factors will depend on the nature of the project. In certain cases, some of the above may not apply at all.

In preparing the pre-qualification document care should be taken to ensure that the questionnaire is clearly worded, contains appropriate questions and that the information requested is capable of being positively assessed.

Having prepared a pre-qualification questionnaire contractors on the long list are invited to respond within a set time scale. Much of the requested information is standard and 7 to 14 days should normally be adequate. In cases where comprehensive information is required, perhaps involving joint venture or complex subcontract agreements, then a longer period of up to 28 days may be necessary.

Replies should be logged, acknowledged and checked for compliance. Those organizations not providing all of the requested information or not responding by the due date should not be surprised if they are not selected to appear on the resulting short list.

9.5 EVALUATION CRITERIA

Evaluation criteria should be established which suit the particular needs of the project. The following may be regarded as typical evaluation criteria for the assessment of pre-qualification submissions.

9.5.1 Technical and professional competence

The contractor should be able to demonstrate that it has the technical capabilities required for the work including, where appropriate, design capability and that it complies with the stated qualifying requirements (e.g. inclusion on an approved list, members of professional/trade organizations, etc.)

Inclusion on an approved list means that the company has given details previously which have been considered satisfactory for inclusion on that list. In some cases the procurement rules of statutory bodies will require inclusion on approved lists which will automatically override the necessity to provide certain basic details; the EC has such a requirement in its rules which allows contractors to gain details and express an interest in a project.

Membership of professional or trade organizations means that a company is affiliated to that organization and operates under its guidance.

9.5.2 Financial stability

The financial strength of an organization provides a guide as to whether it will be able to complete contracts once commenced. The extent of financial/commercial vetting of a company will depend upon the value of work for which it is being considered. The value of the work will also, of necessity, limit the time spent on evaluation.

Financial and commercial data of a company should be requested, including annual reports and accounts. An analysis of what may be contained in these reports and accounts follows. Alternatively, information and data concerning a company can be obtained from business information suppliers such as ICC and Dunn & Bradstreet, who analyse and assess company accounts and report and comment upon the company's financial/commercial credibility.

In addition to these sources of information, data specific to the project can be requested.

9.5.3 Experience

Experience of similar projects

Experience of similar projects means the company has gone through the learning curve and tackled some of the problems that are likely to occur. However, caution must be exercised to ensure that companies which have the necessary skills, experience and adaptability, but which do not have experience of work precisely like that of the proposed project are not immediately rejected.

This broader outlook cannot be taken to extremes, and where there are particular specialities which are critical to the success of the project then the firms invited must obviously be limited to those who have the requisite skills. It should, however, be remembered that an alternative approach may be to allow subcontracting of some specialist activities if the prospective main contractor stands out as being exceptionally experienced, or has a good track record, or is highly qualified in other areas.

The extent and nature of each firm's involvement in these areas of experience is important. For example:

*   If the firm was involved in a joint venture, was it as a principal or as a subsidiary?
*   Did the firm have primary responsibility, or a supporting role for a particular aspect?
*   Was the experience quoted that of the firm, or of an individual gained before joining the firm?
*   Was the project successful/unsuccessful and why?

Auditing and checking of experience on similar projects should be undertaken to establish its validity together with the success, or otherwise, of the project and general project history.

Projects for similar owners

It is useful for firms to be aware of the normal requirements of owners in a particular field, or with a particular background, but it must always be remembered that no two projects are identical and that a modicum of research and consultation will often give the same end result. As with experience of similar projects, this consideration must not be allowed to dominate to the extent that suitable candidates are excluded when they are qualified in other ways.

Track record in the type of services under consideration

In specialist areas such as petrochemicals, pharmaceuticals, nuclear engineering and the like, the kind of service under consideration is often unique and not all contractors/subcontractors will have knowledge and experience of providing that service. Care must again be exercised not to exclude firms who may prove in the long run to be the best by too closely defining the 'type of service'.

Key personnel

Curricula vitae of key personnel may be requested, together with an organigram of a typical management team in order to judge a contractor's management philosophy.

9.5.4 Capacity/resources

The company should be able to manage the manpower and plant requirements for the project. If they have already committed their resources and experienced supervision to other projects it would be unwise to consider them further. Even if they were able to carry out the work by using resources other than those directly owned by them, such as by subcontracting, it may be that their management, planning and coordination resources would be stretched to an unacceptable level and the project suffer.

Company management structures should be studied to ensure adequate reporting relationships between senior supervisory personnel and other levels of management. The availability and depth of experience of permanently employed key supervisory personnel could be crucial to the success of the project. The stability of the company's supervision and workforce should also be taken into consideration.

The company's current and future workload should be investigated to establish that the necessary resources will be available for the project. Overcommitment can cause problems.

The company should be requested to identify those elements of the work that it traditionally subcontracts to other companies.

9.5.5 Quality assurance

The company should be able to demonstrate a commitment to quality management. QA accreditation is one way in which this can be evidenced.

9.5.6 Health and safety

The company should be able to demonstrate a good safety record and have practices and procedure for all employees, i.e. a health and safety policy.

9.5.7 Industrial relations (IR)

Depending on the area and country in which the project is to be executed the company's industrial relations policy could be of significant importance. The company should be able to demonstrate adequate IR practices and procedures and a good track record of successful working relationships with local trades unions.

9.5.8 Project specific

The 'learning curve' required by a contractor who is faced with unusual site conditions or project requirements is very steep and can have a significant effect on performance. There may be a number of such project specific factors which may include:

- unusual construction techniques or processes, e.g. refinery shutdowns;
- unusual contractual relationships, e.g. alliance or partnering arrangements;
- the requirement to work in hazardous areas, e.g. in a live plant;
- unusual geographical areas.

Criteria	Weighting factor	Submission nos./names:									
		A		B		C		D		E	
		Score	Weighted score	Score	Weighted score	Score	Weighted score	Score	Weighted score	Score	Weighted score
Company information	5	10	50	10	50	9	45	9	45	10	50
Technical & professional competence	10	10	100	10	100	8	80	7	70	8	80
Financial stability	10	5	50	6	60	8	80	6	60	8	80
Experience	15	6	90	8	120	7	105	7	105	6	90
Capacity/resources	10	8	80	5	50	7	70	7	70	7	70
Quality systems	10	7	70	6	60	6	60	6	60	6	60
Health & safety	15	8	120	7	105	5	75	6	90	6	90
Industrial relations	10	7	70	8	80	7	70	7	70	6	60
Project specfic	15	5	75	6	90	7	105	6	90	5	75
Notes: Score 1 to 10 where 1–2 very poor 3–4 poor 5–6 adequate 7–8 good 9–10 very good Maximum =1000											
	100	66	705	66	715	64	690	61	660	62	655

Project: Assessor:
Date:

Figure 9.1 Pre-qualification matrix

Climatic, geographical and cultural experience in some parts of the world where working practices differ from those in the UK can have a significant impact on project success. The significance of cultural differences can be reduced depending upon the degree of contact and cooperation with local nationals. Failure to take account of cultural differences can have serious consequences. A company with a local subsidiary or partners in a country together with local plant, yard facilities, etc. would be a distinct advantage. A company offering key personnel with previous experience of working in the area of the project, with knowledge of the language and local culture would also have an advantage.

9.6 SHORTLISTING

Evaluation of pre-qualification submissions is usually undertaken by a team, each member being invited to complete an evaluation matrix. Each of the selection criteria is weighted according to its perceived importance and each submission 'scored' against each criterion. The products of the scores and weighting factors are summated to give each submission a total score. A typical evaluation matrix may be found in Figure 9.1.

The code of practice for competitive tendering makes suggestions about the number of organizations that should be included on a short list and invited to tender. In general fewer than four tenders is insufficient to ensure adequate competition, while more than seven is unfair to tenderers and assessors alike. The choice between four and seven depends upon a number of factors including:

- anticipated project value;
- the cost of tendering – on complex major tenders and large lump sum design and construct contracts the cost of tendering is very significant and it is unreasonable to propose a long tender list. In such circumstances more information should be requested during the pre-qualification period in order to shorten the tender list;
- the degree of complexity and opportunities for submitting alternative proposals;
- the scores from the pre-qualification assessment.

Thus, the highest four, five, six or seven organizations from the selection matrix are advised that their pre-qualification submission was well received and asked to reconfirm their interest in tendering.

9.7 EXAMINATION OF COMPANY REPORTS AND ACCOUNTS

The subtleties of accounting practice make the examination of company accounts a specialist task and what follows can only be regarded as a general overview of this 'art'. The main contents of the report and accounts are described here but do not forget to read the notes to the accounts, which can contain useful snippets of information. It should be noted when assessing company reports and accounts that these are historical documents and are normally one year out of date. It is also worth noting that when comparing data from different years changes in accountancy practice could have had a significant effect.

9.7.1 Directors' report

The 1985 Companies Act requires companies to include in the directors' report, *inter alia*, the following:

- principal activities of the company;
- a 'fair review' of the development of the business during the year with an indication of likely future developments and research and development activities;
- names of the directors and their interests;
- changes in fixed assets;
- details of the company's acquisition of its own shares;
- important events affecting the company since the end of the financial year.

9.7.2 Chairman's statement

This is not a legal requirement and where supplied will usually contain information relating to:

- the general economic climate and future outlook;
- performance of the company in general terms and its future prospects;
- strategy and plans for the future.

9.7.3 Auditor's report

A company is required by law to have its accounts audited and the auditor's statement can provide useful indicators of possible irregularities or company weaknesses. There are some commonly used phrases which are used to indicate specific situations without openly disclosing particular concerns; for example, if there is a reference to the accounts being prepared on a 'going concern' basis this could mean that continuing financial support has not yet been guaranteed.

In any case, if the auditor's report is qualified in any way then further examination is required.

9.7.4 Profit and loss account

From this statement some useful data can be gleaned such as:

- gross and net profit margins;
- unusual increases in expenditure in specific areas (such as administration);
- interest payable on loans and overdrafts;
- directors' and employees' emoluments;
- dividends paid;
- earnings per share.

9.7.5 Balance sheet

This is a snapshot of the company's assets at the end of the financial year and will include details of:

- fixed and current assets;
- creditors falling due within one year (current liabilities);
- working capital (current assets less current liabilities);
- capital employed;
- creditors falling due after one year;
- provisions for liabilities;
- capital and reserves;
- retained profits.

In addition to noting the factual data contained in the annual report and accounts the information can be used to establish ratios which, in turn, allow comparison of the company's performance with that of other companies. It is also useful to calculate the ratios for the previous year to establish trends. There are three key areas which give indications of the strength and stability of a company and for which standard ratios can be established; these are profitability, liquidity and solvency.

9.7.6 Profitability

The basic indicator of profitability is the 'return on capital employed', which is calculated by dividing trading profit by capital employed and expressing it as a percentage.

Other indicators which show strength or weakness in this area are 'profit margin', which is trading profit divided by turnover as a percentage and 'asset turnover', which is the ratio of turnover to capital employed.

A final indicator concerned with profitability is 'trading profit per employee' which is calculated by dividing the trading profit by the number of employees.

9.7.7 Liquidity

The key ratios here are collection period; stock turnover rate; working capital and sales; current ratio; and acid test ratio.

'Collection period' is an expression of the speed with which accounts are paid and is equal to the value of trade debtors divided by turnover times 365 days. 'Stock turnover rate' is a measure of the level of stock needed for a given turnover and is calculated by dividing the cost of goods sold by closing stocks. A high stock turnover rate is desirable as this indicates that stock levels, and therefore currently dead capital investment, can be kept low.

'Working capital to sales ratio' gives an indication of the probable additional cash which must be obtained if turnover is increased. This ratio is equal to working capital divided by turnover. If the ratio year on year is falling this indicates that the company may be overstretching itself in its expansion plans. The result can be failure to pay short-term creditors in order to free cash for expansion. This is a quick route to financial disaster and is to be avoided at all costs.

'Current ratio' is the ratio of current assets to current liabilities. A ratio of more than unity indicates a surplus of assets over liabilities. A company with a current ratio of 1.5 is generally considered to be a good credit risk and therefore fairly secure.

'Acid test ratio' is similar to the current ratio except that it disregards assets which cannot be readily converted into cash such as stocks. The acid test shows what the result would be if the company had to settle with creditors and debtors immediately. If the ratio is less than unity this would not be possible, a low and declining ratio often indicates a rising overdraft; the question to be asked then is whether the company's bankers are likely to be happy.

There is no 'normal' level for either 'current ratio' or 'acid test ratio' due to the number of variables involved. The best use of these ratios is to establish a normal level for the company and look for a recent drop below this norm which would suggest decreasing stability.

9.7.8 Solvency

Solvency relates to the ability of a company to meet its longer term commitments as opposed to liquidity which relates to short-term commitments. In trying to establish the level of solvency of a company two ratios are normally calculated, namely 'gearing' and 'interest cover'.

'Gearing' is a measure of the borrowing levels of the company and can be expressed as the ratio of long- and short-term loans to the total capital employed in percentage terms. Borrowings of up to 50% of capital employed are normally acceptable in the engineering industry.

If gearing is high it means that the company is at the very least close to its total borrowing limits and that it may even be overcommitted.

'Interest cover' is a measure of trading profit against interest charges. If the ratio is less than unity this means that interest charges exceed profits and therefore if the charges were to be met then further capital would need to be raised, for example, by sale of assets or by a rights issue. In practice, this situation would suggest that the profitability of the company is so low that it is really pointless to continue trading unless the award of further guaranteed profit-making contracts is imminent.

9.7.9 Financial/commercial capacity

In addition to general published company information specific data relating to the project should also be obtained.

Financial limits

The company should be able to undertake the financial investment required for the project. Financial resources should not be stretched to the limit either by the size of a particular project or by other projects that the contractor is already committed to.

Banking details

Reference from the company's bank should be sought to establish the current up-to-date financial situation and lending facilities available.

Insurances

The company should give details of its insurance cover in connection with employers' liability, public liability and professional indemnity to establish and confirm its suitability and capacity to undertake the work or to see if alterations to the levels of cover are required.

9.8 PRE-QUALIFICATION INTERVIEWS

Depending on the proposed number of tenderers to be included on the final tender list a selection of the most capable, enthusiastic and experienced contractors/subcontractors, as determined by the pre-selection process, may be invited for interview.

The purposes of the interview are to clarify the information that has been provided, confirm the contractor's/subcontractors' continued enthusiasm and ability to undertake the project and to meet the personnel who are likely to work on the project.

On some projects a pre-qualification interview may not be required or desirable due to size, time or cost restraints and any interview will be held over until after tender.

9.9 CONTRACTING STRATEGY

Contracting strategy involves two main decisions, how to break up the overall project into appropriate, separately tendered, packages of work and the type of contract to be used.

The choice of pricing strategy involves a number of factors including:

- What date will the various parts of the work have to be awarded in order to maintain the programme?
- What design information will be available at tender enquiry, at award and during construction – will this preclude a lump sum contract or will design hold up construction?
- What is the likelihood of change?
- What is the potential for disruption?
- What degree of control is required over the manner of execution by the owner or his management contractor?
- What risk is the owner prepared to accept?

Although the contract strategy will be established early in the project and incorporated into a contract plan as described in section 2.6.9 on management of engineering projects, it is nevertheless dynamic and should be kept under review and amended to accommodate the changing circumstances in the project.

9.9.1 Tendering risk and control

Many of the above factors relate to design status and the likelihood of change or disruption to the work. A turnkey contract offers the lowest financial risk to the owner in that the final cost can be guaranteed but it requires that definitive information is available with the enquiry and it does not provide the owner, or his project team, with the ability to control or have continuing input to the work. By comparison the final cost of a reimbursable contract cannot be guaranteed but the owner can control the work and make changes without major contractual risks.

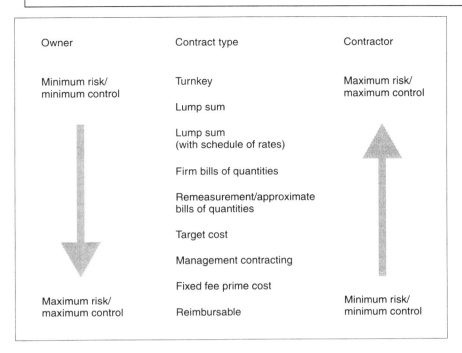

Figure 9.2 Relative risks amd controls

In order of minimum to maximum owner risk/owner control the types of contract commonly used and the relative risks and controls are shown in Figure 9.2.

9.9.2 Types of contract

The type of contract to be used depends on a number of factors, including:

- availability of design information at enquiry and award: potential for change;
- potential for disruption or delay;

A project could include several main contracts each let on a different basis as may be appropriate. Sometimes it is advantageous to let different portions of one project on a different basis, perhaps obtaining fixed prices for designed elements and other appropriate methods of payment for elements yet to be designed.

The type of contract to be used can also have the potential to affect the programme due to the way in which design can overlap the construction phase. An example of this is given in Figure 9.3.

The types of contract and the circumstances under which they may be used include the following.

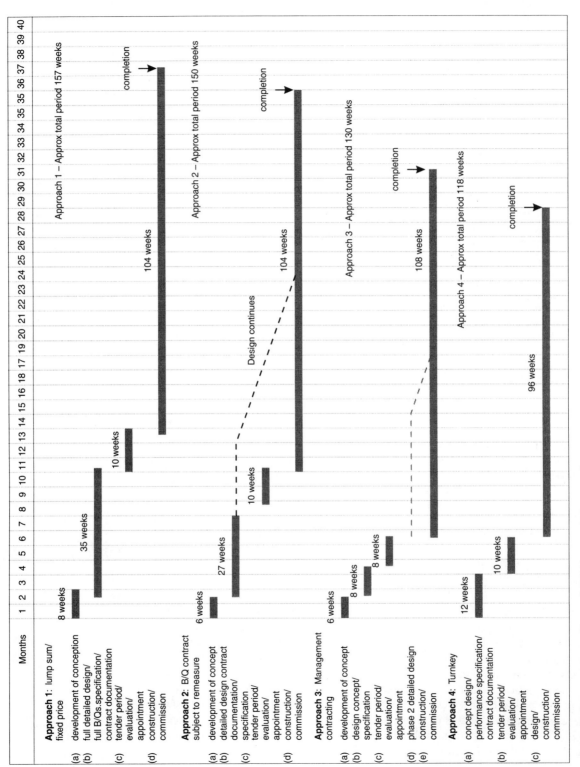

Figure 9.3 Programme effects of differing tender approaches

Turnkey

A broad scope of work plus all process and performance requirements is required with the enquiry and a fixed lump sum tendered for the design, purchasing, construction and supervision of the work, etc., through to commissioning and performance tests.

Lump sum

All relevant design information is totally completed prior to it being handed over to the contractor who quotes a 'lump sum' for carrying out the works. A schedule of rates may be included in the enquiry to evaluate unavoidable change or extra work.

Lump sum (with schedule of rates)

As above but although the relevant design information will be substantially complete a schedule of rates or approximate bills of quantities may be provided to cover items for which design is incomplete for later evaluation.

Firm bill of quantities

Design fully or substantially complete at enquiry stage. The contractor is required to tender against a detailed, itemized pre-measurement of the work issued with the enquiry. Work is not remeasured unless changed.

Remeasurement/approximate bills of quantities

Design incomplete at enquiry stage and completed as the work progresses. A schedule of rates or approximate bills of quantities is issued to the contractor to obtain rates to be applied to a remeasurement of the works undertaken as the work progresses.

Target cost

Design incomplete at enquiry stage and completed as the work progresses. The work is commenced against a 'not to exceed' budget with incentives for working within or below budget and perhaps for early completion.

Management contracting

All preliminary or conceptual design information and criteria relating to performance requirements is handed over to the management contractor who manages the project from start to finish. The management contractor, who may be employed on a lump sum or reimbursable basis, arranges discipline tender packages and supervises work to completion. Contracts for supply of equipment, materials or construction work may be placed by the management contractor for and on behalf of the owner, or alternatively, may be directly subcontracted by the management contractor. In this latter case the construction costs would usually be reimbursed to the management contractor.

Fixed fee prime cost

Design incomplete at enquiry stage and completed as the work progresses. A form of part reimbursable contract with for example the contractor's main office, overheads, profit, etc., contained in a 'fixed fee' and the remainder, i.e. labour, supervision, site establishment, materials, etc. being reimbursed at cost.

Reimbursable

Design incomplete at enquiry stage and completed as the work progresses. Costs are fully reimbursable with the contractors having a mark-up or lump sum(s) for profit and overhead, etc.

9.10 TENDERING PROCESS

9.10.1 Objectives

The objectives of the tendering process are to ensure that:

- the owner receives the most favourable price/tender;
- bona fide competitive tenders are received;
- tenderers have the necessary information to properly price the work and are given an equal commercial opportunity;
- truly competitive tenders are obtained in a uniform and directly comparable basis;
- confidentiality is maintained;
- tender evaluation is fair and efficient;
- all concerned are aware of the procedures and programme leading to contract placement.

9.10.2 Tender documentation

The tender document, issued to those on the tender list, must be the same for each tenderer and should contain all the information necessary for the tenderer to appreciate the full extent and content of the work. The tenderer must be given sufficient time to analyse, complete and return the tender by the stated time with all the information requested.

Tender documents vary in their content according to the scope of work, type of contract, form of agreement and selection criteria. The following may be regarded as typical of the documents prepared by the procurement team:

1. covering letter;
2. introduction;

3. instructions to tenderers;
4. form of tender;
5. general conditions of contract and form of agreement;
6. special conditions of contract;
7. specification:
   - scope of work;
   - general;
   - technical including drawing package;
   - site conditions and regulations;
   - safety and fire regulations;
8. project master schedule;
9. contract sum analysis/schedule of rates/bills of quantities;
10. forms for bonds/parent company guarantees, etc.

The following are typical of the documents to be prepared by the tenderer as appendices to the tender:

- method statement;
- programme, cash flow, labour histogram;
- management structure;
- curricula vitae for key personnel;
- track record information;
- safety plan;
- quality plan;
- proposed subcontractors or joint venturers;
- any design information required;
- update of new contracts awarded or bid since pre-qualification.

If the tender list has been produced following a pre-qualification exercise it would be reasonable to suggest that tenderers should not be required to resubmit information previously submitted as part of the pre-qualification submission. In such circumstances tenderers may be requested to confirm that their previous submission was still valid or otherwise to provide further information in the event that the previous submission did not fully address the current requirements or the contractor's circumstances have changed. This last point is particularly valid in connection with current and anticipated workload, etc.

9.10.3 Tender period

It is important that tenderers are given sufficient time to formulate their bids. Insufficient time will either make the tenderers decline to bid, or to include larger risk costs. In deciding the tender period consideration should be given to the amount of design, if any, that the tenderer may have to undertake in order to complete his tender and to the fact the tenderer may have to issue and receive tenders from his subcontractors prior to finalizing his quotation.

9.11 TYPES OF TENDER

9.11.1 Open tenders

Open tenders are utilized when obtaining prices for small elements of work, materials and equipment where there is no necessity for confidentiality. Tenders are requested with a 'guide' date for their return. It is normal to include address labels, but purely for the convenience of returning the tenders to the interested party.

There is no official opening as tenders may be received before, on or after the 'guide' date. Tenders, however, should be logged in and a brief analysis/comparison completed on receipt of all the tenders. This analysis should be placed on file for audit purposes.

9.11.2 Sealed tenders

This is the 'formal' method of obtaining tenders. The date and time for receipt of tenders is fixed. Thus any tender received after the prescribed date and time may be disqualified. Envelopes with labels in place are usually included with the tender package for return of tenders to aid the identification of its importance and confidentiality. Some owners require the financial section of the proposal to be returned in a separate sealed envelope.

All questions raised by the tenderers during the tender period should be channelled through one person, usually the contract engineer/quantity surveyor/buyer who will pass them on to the relevant parties. Once the answers are formulated the questions and answers should be distributed to all the tenderers by the contract engineer/quantity surveyor/buyer to ensure information is common and hence like for like tenders received. Care should be exercised to avoid passing on any innovative design information to other tenderers.

Tenders, once received, are logged in by recording the date and time of receipt.

At the date and time for receipt of tender, all tenders are opened and each page containing prices and/or rates should be signed or initialled by all the members of the tender opening committee which usually comprises as a minimum two people, the manager responsible for the work and the contract engineer/quantity surveyor/buyer. A record is made of the opening, those present, tenders received, and their respective amounts.

9.12 EC DIRECTIVES

The European Union (EU) may be regarded as a political body that uses the European Commission (EC) as an executive body. Thus the EC issues directives on behalf of the EU. The EC directives aim to open up trade across Europe for suppliers and allow purchasers to realize potential savings by the introduction of fair competition.

The directives and their embodiment into UK law (as at October 1995) are as follows (values are originally in ECUs therefore sterling values are approximate):

- the Utilities Directive 90/531/EEC, SI 92/3279 covering supplies and works contracts for operators in the energy, water, transport and telecommunications sectors. Threshold for supplies and services is £300 000 (except for telecoms sector £450 000) and for works £3.7 million;
- the Supplies Directive 93/36/EEC; SI 95/201 for the purchase of goods with a value of £150 000 or more;
- the Works Directive 93/37/EEC; SI 91/2680 for construction and engineering contract with a value of £3.75 million or more;
- the Services Directive 92/50/EEC; SI 93/3228 for various services including consultancy with a value of £150 000 or more;
- the Compliance Directive 89/665/EEC sets out procedures for objection and the various measures and penalties available for enforcement.

Note: There is no separate Statutory Instrument for the Compliance Directive as its provisions are built into the other Statutory Instruments.

Failure to comply with these rules can result in a challenge by an unsuccessful tenderer who can refer the matter to the High Court (the Court of Session in Scotland). The High Court cannot set aside a contract which has already been awarded, but it can award damages. Complainants do not have to prove that they would have won the contract if the correct procedures had been followed, they only have to show that a contracting entity did not comply with the rules.

Infringements of the rules may be broken down into two categories – procedural infringements and discrimination. Procedural infringements include administration errors such as failure to comply with the laid down time limits, incorrect notices, etc. Discrimination covers technical specification, operating requirements, financial specifications or other actions which discriminate in favour of local or national suppliers.

The court determines the level of damages to be paid. These are not restricted to the costs incurred in preparing the tender submission, but can cover lost profits, opportunity, costs, etc.

To comply with the EC directive all requirements for contracts, including services, valued at above £150 000 (£3.7 million for works), must be publicized in advance throughout the member states by means of a Pre-Information Notice (PIN) placed in the Official Journal of the European Community (OJEC). The OJEC, which contains advertisements from all member states, is published daily and is available within the UK from Her Majesty's Stationery Office (HMSO). The rules ban prior parallel advertising within newspapers and trade publications.

All suppliers within the EC must be given an equal opportunity to tender for the business.

Tenders must be invited in accordance with one of the prescribed procedures: open, restricted, negotiated, accelerated or urgent. Special rules apply for concessions and design contests. The open or restricted procedures are the usual ones as there are very tight restrictions on the use of the others. Each procedure imposes minimum time scales covering every tender/procurement activity. These time scales must be met to ensure that the owner is not subject to legal action under the Compliance Directive.

Contracting entities must also comply with regulations covering standards. They must use European standards if available and after that, a hierarchy of alternatives. The essential requirement is that the specification selected may not discriminate against suppliers from other member states.

9.12.1 Contract award criteria

The choices of selection criteria offered by the EC are either:

*   the lowest price or
*   the most economically advantageous.

While lowest price may be adequate for relatively low technology, proprietary items or low risk purchases, it is almost certain, that for major purchases, owners will use most economically advantageous (MEA) as the basic contract award criteria. Factors such as the following may be appropriate:

*   delivery or completion date;
*   running costs or overall cost effectiveness;
*   quality and reliability; aesthetics and functionality;
*   technical/design capability and levels of technical assistance offered;
*   after sales service and commitment to spare parts;
*   strategic security of supplies;
*   price.

9.13 RECEIPT AND ANALYSIS OF TENDERS

9.13.1 Analysis of tenders

Although variable for each procurement exercise, the technical submission and the commercial submission are normally evaluated separately and without consultation with the assessors of the other submission. Some owners require that only those technical submissions achieving a particular threshold have their accompanying commercial submissions opened, all others being returned unopened to the tenderers.

Many of the criteria chosen for selection of a contractor and questions asked of a tenderer may be subjective, resulting in difficulty in agreeing the 'scoring' of the tender submission. In order to assist in the evaluation of subjective criteria, the evaluation team may find it beneficial not only to define and agree the marking criteria prior to receipt of tenders, but also to state the objective behind asking the question and to stipulate the information required.

For example, objective/requirement on workload may be:

- objective: to provide to the owner such information as will demonstrate that the tenderer will have available such facilities, plant and equipment and experienced personnel as are necessary to undertake the work during the contract period identified on the project master schedule;
- requirement: the tenderer shall provide details of his future commitments, both for himself and any proposed joint venture partners and subcontractors.

9.13.2 Technical appraisal

It is normal when evaluating competitive tenders to draw up comparison schedules (technical and commercial), where the particulars from all tenders being reviewed can be presented in a tabular form. Any outstanding variation from the normal or expected should be examined and commented on.

During the detailed examination, some gaps in information provided or anomalies within the tender may be discovered which, time permitting, will have to be clarified with the tenderer.

If unit prices are tendered a comparison of the prices of items should be made and any omissions or anomalies noted. While the apparent technical details may not reveal any technical shortcoming, a wide price difference may indicate that the tenderer is not meeting the expected technical standard, has not understood the work content or has simply made an error.

In some situations a points system may be used for marking and ranking tenders in order of technical merit.

The points system is useful where no prices are available to the technical evaluation team, or in the case where technical ability is the overriding criterion.

9.13.3 Commercial and financial appraisal

An arithmetical check should be made of all tenders and a detailed analysis of each tender carried out.

A price comparison table should be drawn up using the information presented in the tenders (i.e. schedule of rates, bills of quantities, etc.). If the enquiry is in the form of a schedule of rates then estimated quantities should be set against each item and an estimated final sum calculated.

It is at this stage that early signals of variations to the client's preliminary budget may be identified and projected. It may be necessary to consider actions to reduce costs through design or specification changes or to highlight the need to exercise particular cost control to avoid or reduce projected cost increases.

A major concern of the tender appraisal is to ascertain whether the tenderer has understood and priced all the work described in the tender documents and whether inconsistently high or low rates may have a disproportionate effect on the final sum to be paid to the contractor.

A cost model of the project can be extremely valuable to the tender analysis team. Such a model can be set up on a spreadsheet and risk assessments can be made by adjusting key cost values to simulate potential claims, scope changes, daywork allowances, etc.

The cost model can also allow the analysis of anomalies or omissions in the tender and to estimate the cost of the omitted elements.

Decisions on the basis of bid price can also be enhanced by a time based view as cash flow can affect the viability of the project or, for example, an early finish can improve the financial success of a project despite a higher initial capital cost.

In the event of a tenderer making an error in a fixed price tender it is good practice to invite the tenderer either to stand by his original tender or to withdraw the tender but not to invite a correction to be made.

9.13.4 Alignment of tenders

On large projects where the information provided with the tender enquiry does not precisely define the work in all respects, it is inevitable that differences will arise in detail design, supply and services proposed by the tenderers.

In order to make a direct comparison of tenders an adjustment to some tenders will be necessary to align them with a norm for evaluation purposes. This norm need not be the final contractual scope but it avoids later adjustments if it is.

The alignment process can be carried out by the evaluation team themselves making estimates or by reference to the tenderers requesting additional information. If the situation is simple and the estimated cost effect is low, it is quicker for the evaluation team to make the assessment themselves. Where the situation is complex and large costs involved, the tenderers should be consulted. If this situation can be foreseen prior to inviting tenders then a two stage form of tendering may be written into the tendering procedure to cover this contingency.

Tenderers may seek to qualify their tender and the evaluation team should consider each statement made by the tenderer, anticipate why it was made and either seek withdrawal of the qualification or evaluate its impact on the work.

If it proves necessary to approach the tenderers for alignment purposes, all questions from all team members should be compiled into one document for each tenderer. The questions should be unambiguous and preferably in a form that can be answered by 'Yes' or 'No' or by a quantitative statement.

The evaluation team should indicate the additions or deductions for alignment purposes which they deduced from the information submitted by the tenderer and/or by their own assessment.

9.13.5 Key personnel

Curricula vitae (CVs) of all the proposed key personnel should be acquired along with an organigram of the proposed team. Key personnel should, of course, be the personnel who are actually going to undertake the project and not be marketing executives, front men, etc. who have good CVs but will not be part of the team. Assessments are often made by a combination of desk top studies, the taking up of references and interviews.

9.13.6 Capacity in terms of manpower

The company should be able to handle the manpower investment required for the project. If such matters have not been reviewed during pre-qualification they should be considered at this time (see section 9.5.5).

9.13.7 Current and future workload

The company's current and future workload should be investigated to establish that the necessary resources will be available for the project. Overcommitment can cause problems. Also the company's current projects should be investigated to ensure that any delays in completion will not impinge on the subject programme.

9.13.8 Proposed subcontractors and suppliers

The subcontractors and suppliers proposed for the project by the company should be checked to ensure that there is no conflict with the owner's preferred list.

9.13.9 Project execution

The company should be able to demonstrate an understanding of the scope, content and any logistical and technical difficulties associated with the work by submitting with the proposal a description of the sequence and method of construction, together with details of the proposed organization. The company may also be requested to provide a labour histogram and details of proposed project controls.

9.13.10 Programme

In many contracts the completion date is an obligation but the programme to achieve it is not. A proposed programme is usually required to be submitted with the tender establishing key dates, phases, durations, etc. for the proposed works. This is the key to establishing project controls as well as demonstrating the tenderer's understanding of the project. The assessor would need to check this outline programme to see that it generally meets with project requirements, delivery dates of material and equipment and the work of others.

9.13.11 Post-tender interview

Once the preferred tender has been established (which may not necessarily be the lowest priced tender), discussions will be held with that tenderer to resolve any outstanding anomalies, commercial and technical, that may remain. This post-tender interview is designed to clarify any points on the contractor's/subcontractor's submitted tender, confirm acceptance and enthusiasm to undertake the project and meet the proposed project team who should be the team who are actually going to undertake the project. Minutes of the above discussions should be taken which may be included in the contract documentation.

If important anomalies cannot be resolved satisfactorily, then discussion may commence with the next favoured tenderer. The resolving of the outstanding anomalies allows the contract or purchase order to be awarded.

9.13.12 Tender evaluation report

The tender evaluation team should always present a tender evaluation report to the project manager that clearly states the recommended contractor and the reasons for the recommendation. Whether the review criteria appear in either the pre-qualification (if there is one) or in the tender assessment depends on many factors. Some will obviously only relate to the tender assessment but others may be relevant to one or to both.

9.14 FINALIZATION OF CONTRACT DOCUMENTS

Once approval has been obtained to the award of a contract from the various project and corporate managers, co-venturers, etc., the contract documents will be assembled and listed.

Depending on the number of modifications and amendments made to the original tender enquiry, the contract may comprise the priced tender document, together with reference to, or copies of, all documents recording such modification/amendments. However, a heavily amended contract can be difficult to interpret and use and it is frequently preferred that a 'clean' copy of contract is produced, in which the modifications/amendments are incorporated into the body of the document.

9.15 CONCLUSION

It can be seen from the above that detailed evaluation of potential tenderers is a complex process of great value if tackled properly. It depends to a great extent upon the experience of those performing the task, not only in the evaluation itself but in deciding what factors to consider and allocating relative importance to each.

It is probably true to say that if two teams of people were to evaluate a set of companies they would produce two different lists, if not in respect of the names included then certainly in the order of preference. Advance preparation is essential if a correct assessment is to be made and adequate auditable records of the decision making process produced and maintained. The use of a compliancy matrix and evaluation matrix is highly desirable and the items listed should be given a relative weighting if possible, but at the least should be classified as 'essential' or 'desirable'.

The stability of any structure, whether it be physical or organizational, depends on a firm foundation. In terms of project management the foundation of a successful implementation phase is the selection of the correct contractor. To achieve this, it is essential to carry out a thorough evaluation within a structured selection process.

Procurement
of Materials
and Equipment

Downshaft pumps in
water ring main, London,
England.

Owner Thames Water Plc.

10.1 INTRODUCTION

This chapter addresses the procurement of materials and equipment. Procurement of construction contractors and subcontractors are covered in Chapter 9.

10.2 SCOPE OF PROCUREMENT ACTIVITIES

Throughout all industries there are five guiding principles that govern the procurement function: they are known as the 'Five Rs':

1. right quantity;
2. right quality;
3. right supplier;
4. right price;
5   right time.

To ensure compliance with these principles, the procurement function normally comprises the activities relating to purchasing, expediting, inspection and shipping. In addition to these activities is material control, which is considered in Chapter 5, Part 4 on material control. The procurement function is responsible for ensuring that all equipment and materials, together with associated drawings, documents and data are delivered in such time as will enable the engineering, construction and commissioning programme to be met and subsequent plant operations carried out.

10.3 REQUISITIONS

A requisition is a statement of requirement prepared by a user department (usually the engineering or construction department) which is used as the basis against which materials and equipment on a project are purchased. Requisitions will contain information regarding scope of supply, quantity, specification and final location within the work, together with details of any special requirements. Required delivery dates to meet the construction schedule will be attached by the planning department.

10.4 THE PROCUREMENT CYCLE

The term 'procurement cycle' is frequently used to describe the time taken from the identification of a material or equipment requirement to the placement of a purchase order. The cycle is established at the planning stage of a project and delivery periods estimated, so that the programme of requisition production can be formulated to meet site delivery requirements.

Drawings and data from a supplier may be critical to the design; delivery dates of the hardware may not therefore be the only controlling factor.

Activities within the procurement cycle are normally as follows:

- receipt of enquiry requisition (by procurement);
- pre-qualification of bidders;
- issue of enquiry;
- receipt of tenders;
- shortlisting of tenderers;
- commercial and technical analysis of tenders;
- preparation of bid summary;
- receipt of purchase requisition (by procurement);
- placement of purchase order.

Time taken for each of the above activities depends on the complexity of the equipment and materials to be purchased, however, standard times are agreed for planning purposes.

Standard times will run from completion of the previous activity and may be based on a typical enquiry for a packaged equipment item. For bulk materials, times will be reduced considerably, and for major or complex equipment items, times will increase, especially when the supplier is required to carry out design work to comply with tender requirements.

The procurement role on a project starts at the commencement of the detailed planning stage. The procurement cycle and individual delivery periods will be incorporated into the planning schedule, to ensure that design information and the materials or equipment items are available to meet design and construction requirements. This exercise not only checks the validity of the overall programme (design and construction), but also enables planning of required procurement manpower to take place.

10.5 THE PROCUREMENT TEAM

For major projects it is common to have a dedicated procurement team. The manager of the team will report directly to the project manager. Besides his procurement responsibilities the procurement manager will be part of the project management team. All procurement matters on a project will be dealt with by the procurement manager and he will liaise directly with the owner's project manager or procurement coordinator (depending on size of project) on all subjects concerning the purchase of materials and equipment.

The general responsibilities of the procurement team are as follows:

- purchasing: identification and evaluation of potential suppliers, enquiry issue, tender receipt and analysis, order placement/administration/close-out of the order;
- expediting: ensuring the timely receipt of design data and materials/equipment to meet design/construction requirements;
- inspection: ensuring that materials/equipment ordered are to the correct specification and meet the required quality;

- shipping: ensuring that materials/equipment released by inspection are shipped in a timely and cost effective manner to the correct site; also responsible for maintaining import records and liaison with customs authorities.

10.6 EU REGULATIONS

The EU directives that cover purchasing in the oil, gas and certain other industries are enshrined in UK legislation through the Utilities Supply and Works Contracts Regulations 1992. This legislation covers the method of establishing a supplier qualification system and how the purchasing process is governed to ensure full and fair opportunity for companies from all EU countries. The rules cover the advertising of requirements, selection methods for tenderers, method of enquiry, evaluation criteria, award notification and required records. Reference should be made to Chapter 9 on contractor/subcontractor selection where the major aspects of this legislation are outlined.

Procurement work has increased due to the implementation of this legislation. The legislation is both complex and vague, which has resulted in a variety of approaches being adopted by different companies. However, it is important to note that the legislation only covers utilities (oil, gas and water companies) and not contractors purchasing in their own names.

10.7 COMMUNICATION

The entire relationship forged by a purchase order between purchaser and supplier is built on good communication. Where communication links break down, errors occur and strains in this relationship take place. Communication is obviously a two-way process and efficiency will be enhanced where goodwill and trust are generated. The objective of good communication is to avoid problems, but if problems do occur, good communication should ensure that they are identified early and have less impact on the work. All parties to an order should be seeking to create a successful long-term relationship and good communication is vital to the achievement of this aim.

10.8 TERMS AND CONDITIONS

10.8.1 Purchase order versus contract

Purchase orders are of course legally binding contracts; many companies, however, refer to the relatively simple documentation required to buy material as a purchase order and the larger, more complex documentation required to place, control and evaluate work of design or construction as a contract. Because of the difference in content and method of administration they are frequently handled by separate sections.

Different companies have different interpretations of what comprises a purchase order and what comprises a contract. The clearest demarcation between the two is that where work is required at a construction site this should be the subject of a contract, with a purchase order being applicable where the arrangement is for supply and delivery only. However, where there is a large design element to the work a contract is frequently used in place of a purchase order, to give added protection to the purchaser. Most companies will have their own policy on this subject, but it should be noted that the decision between purchase order and contract is rarely one of value alone.

10.8.2 Obligations of purchaser and supplier

The purchaser's principal obligations under a purchase order are to define precisely what is required and to pay the price agreed when items are correctly delivered according to those requirements. However, this as stated is a very simplistic approach and the purchaser does have other obligations. Once an order is placed, there may be design documentation for the purchaser to approve according to approval schedules agreed in advance with the supplier, so that delays do not affect delivery dates. For complex or high value orders, progress payments may be agreed so that when specific milestones are reached (either time or event based or both), payments become due. The purchaser also commits to take delivery at specified dates either on an ex-works basis (purchaser arranges collection) or at a construction site.

For the supplier, his obligations can be summarized in three words — quality, quantity, time. By accepting an order, the supplier agrees to provide the specified item(s) to the correct specification and quality, in the correct quantity, and to produce the goods in the agreed time scale. Other obligations such as provision of design data, progress reports, invoices and maintenance of records also exist but are subsidiary to the main three.

Where delivery is made to a construction site, then the stores organization or construction/contractor will take delivery of the items. They will be responsible for off loading and safe storage of items, until required for construction. In order that payments can be made to suppliers, they will ensure that items delivered are correct in number and that no shortages or damage have occurred in transit.

10.8.3 Basis of order

Purchase orders will specify what is needed, to what specification, the quantity required, the price and delivery date and location. The purchase order is confirmation of all that has passed between the purchaser and supplier from a simple telephone call at one end of the spectrum, to several months of detailed negotiations at the other end. A variety of documents may be either referenced in, or attached to, the order and these expand the order's content. Such documents will range from guides to reporting requirements, specifications, etc., to the agreed terms and conditions under which the order is placed. The order from a purchaser represents acceptance of a supplier's offer and without agreed conditions, an enforceable contract may not exist. Such conditions are important to both parties because they ensure that responsibilities are known.

10.8.4 General terms and conditions

Most companies have their own purchase order terms and conditions and all would prefer to trade using their own terms rather than those of another company. However, it is rare in the engineering industry to accept a supplier's terms and conditions even where near monopoly situations occur. Therefore, an important part of the procurement process is the agreement of terms and conditions before order placement. Where the contractor is acting in the owner's name or on a 'for and on behalf of' basis then it is usual for owner's terms and conditions to be used. However a contractor undertaking procurement on its own behalf will frequently use 'in-house' conditions with amendments to reflect any special terms of the main contract.

From a supplier's viewpoint, the important terms of any order relate to warranties, liabilities, penalties and terms of payment. It is usual in the oil and gas industry for extended warranties to be sought, due to the time it takes for projects to be completed, with a warranty period of three years not being unusual. Regarding terms of payment, the standard is frequently a single payment, but milestone payments may be linked to completion of drawings or submission of data where such is required to progress the main design work. Suppliers will often request progress payments where delivery schedules are lengthy or material costs high.

10.8.5 Call-off orders

Call-off orders or supply agreements are used to cover bulk material requirements where final quantities are not known at the date of original order placement. They are used to avoid the necessity of placing several orders for the same material with often the same suppliers. By using such a mechanism, a source of supply can be established early in the project programme, design data obtained, a single standard specification established and administrative costs reduced. Such orders can be placed on a per project basis or for a specific time, covering an owner's requirements on several projects.

When issuing enquiries on this basis, it is important to have a reasonably accurate estimate of total quantities and items required, so that best price and discounts can be obtained. From a supplier's viewpoint, such orders are beneficial as they assist production planning and secure a long-term commitment. Call-off orders are used for piping materials, structural steel, valves, cable, etc. For orders that extend over one year, and where prices are increasing, an escalation formula can be agreed to share the risks between the parties.

Terms and conditions for such orders would not vary to any great degree over those for a one-off requirement, but would address items unique to call-off situations, i.e. no guarantee of final quantities, price escalation formulae, buy-back arrangements, etc.

10.8.6 Bonds and guarantees

A variety of commercial tools are available to a purchaser when drafting terms and conditions, which are used both to provide comfort to the buyer against potential problems and to limit risks to the buyer. Those most commonly used are as follows:

- Parent (or group) company guarantee issued by the ultimate parent of a supplier guaranteeing that its subsidiary company will perform the work in accordance with the order and that if their subsidiary company should default, that the parent company will complete or pay for a third party to complete the work. This type of parent company guarantee is readily provided by suppliers as it does not cost them anything (excluding minor administration costs).
- Bank guarantees, issued by the supplier's bank to cover advance or progress payments and/or early release of retention moneys held. Such guarantees impose a financial liability on the supplier that normally reduces the limit of any credit available to the supplier.
- Retention, withheld from invoices paid (usually 10%). Release of retention is usually on delivery (50%) with the balance paid at the end of the warranty period.
- A performance bond is issued by a bank or insurance company for a certain percentage of the order value (usually 10%) which is called in the event of non-performance by the supplier.

In deciding which of the above tools (if any) to use, the buyer must assess the potential risks of the supplier's failure to perform, against the additional costs incurred by their use. The provision of any of the above is not in a supplier's interest as they cost money to obtain, delay part of his payment or reduce his ability to obtain credit.

The other most common method of reducing the buyer's exposure to risk is the introduction of liquidated damages. Liquidated damages are normally written into purchase orders to provide protection against late delivery; they are normally applied on an escalating basis, typically at a rate of 1% of total order value per week, up to a maximum of 10%. Although widely used in purchase orders, liquidated damages are seldom invoked, and more often than not, the potential threat of imposition ensures that suppliers perform. In today's climate of improved supplier relations, partnering, etc., such threats are seen as part of the historical adversarial approach to suppliers and their use is much reduced.

Insurance matters, including guarantees and bonds, are further considered in Chapter 6.

10.8.7 Warranties

Most if not all work will be subject to a warranty (sometimes called a guarantee) against defective workmanship or materials. The period of the warranty will be stipulated in the various contracts and purchase orders and will depend on the wishes of the owner.

Warranty periods usually run for a period of 12 months from the date on which the whole installation is handed over. Since the construction of a plant can take many years it follows that warranty periods for construction work completed or equipment supplied during the early stages must have long warranty periods.

In the event that a main contractor purchases equipment, etc. in his own name then, subject to the particular contract conditions, any warranty provided by that main contractor will include any materials, equipment, subcontract work, etc., procured under that contract. If, however, an item of equipment, not purchased in the owner's name, is required to have a warranty in excess of that applying to the main contract, unless the main contract makes provision for this additional warranty requirement, it will be necessary to ensure for a clear route to be established through which the benefit of the warranty can be passed from the purchaser to the owner either by assignment of the warranty or by way of a collateral warranty.

10.9 CERTIFICATION AND CERTIFYING AUTHORITIES

To ensure that the correct materials and equipment are delivered, it is normal practice to require suppliers to provide copies of material certificates, type test certificates, etc., to cover the items ordered. Certain documentation may be required so that the final completed plant can obtain the necessary approvals to start production. For UK offshore projects, this involves the employment of a certifying authority. The certifying authority is an independent company that ensures that the design, materials and equipment have been produced to the required levels so that the plant obtains the certificate of fitness from the Department of Trade and Industry.

10.10 MATERIAL TRACEABILITY

Material certificates, type test certificates and the like will be of limited value if the material or equipment to which they apply cannot be identified in the construction site store or in the completed project. Allied to the need for certification from suppliers is the need to be able to maintain material traceability. With bulk materials such as steel plate or pipe, reference numbers equating to those marked on the relevant certification can be 'hard stamped' on the items. Such markings are known as heat numbers and are repeated when the material is cut into smaller pieces. For equipment items, tag numbers will be affixed either on nameplates or to metal tags. Both methods allow ease of traceability once installed so that if there is failure or other problems, the original supplier can be identified and the source of material traced.

10.11 SPARES AND SPECIAL TOOLS

10.11.1 Spares

There are three major types of spares: commissioning, insurance and operating spares.

Commissioning spares are usually purchased either at the same time as the main equipment or separately during the manufacturing phase, but often on the same order. Typically these spares are items that may need immediate replacement during commissioning, i.e. gaskets, bulbs, filters, bolts, etc., which are of low value. Depending on installation requirements, these spares are delivered either with the main equipment, or immediately before commissioning takes place.

Insurance spares are, as their name suggests, purchased to provide insurance against a major problem with a piece of equipment. They are typically purchased where the cost of a major breakdown would lead to an interruption in output, thereby causing serious financial loss. Types of spares in this category are spare rotors for main oil line pumps, spare jibs for platform cranes and spare turbines for either compressor or power generation packages. Typically, such items have lengthy lead times and are often of high value. As with commissioning spares, such items are either purchased at the same time as the main equipment (but with a later delivery date) or on a separate purchase order.

Operating spares are initially for either one or two years' operation; these are traditionally purchased separately from the main equipment and quite often are purchased by the owner direct, rather than by the project procurement team. Orders may be placed later than the main equipment, with deliveries taking place immediately before plant startup. Initial operating spares are only purchased direct from the equipment supplier for the initial requirement, for further ongoing needs items are bought direct from the manufacturers.

The cost of all types of spares should be taken into consideration when making the original supplier selection, as high operating costs can outweigh any initial advantage on initial capital cost. Due to the high cost of spares, both in capital investment and storage expenses, suppliers are now being persuaded to store spares at their cost, rather than expect the owner to purchase and store.

10.11.2 Special tools

Some items of equipment require the use of special tools for the purposes of installation, commissioning or maintenance. The requirement for special tools will be identified within the requisition and incorporated in the enquiry.

10.12 SUPPLIER DATA

Equipment supply packages may have a considerable design content, and the timely receipt of data (drawings, specifications, procedures, certification, etc.) for ongoing design purposes is critical. In some cases, the receipt of such data can be more critical to the overall project programme than the receipt of the equipment itself. When issuing an enquiry, the buyer will specify the data required and the time scale for its receipt. Dates for submission will be agreed with the supplier before order placement, and the supplier will be expedited against these dates. Once data are received, the engineers will be expedited to ensure comments or approvals are made within the agreed periods (typically 10/15 days) and any resubmissions required from the supplier will also be expedited.

Receipt of supplier data must be constantly monitored throughout the project programme to maintain design progress and individual supplier programmes. Final payments should not be made until all data and records are received and approved.

10.13 QUALITY ASSURANCE

All orders placed are subject to a QA audit if so required. Increasingly, more suppliers are qualified to BS 5750 or ISO 9000 and are therefore independently audited against this standard. Quality assurance audits take on average one to two days. They cover the supplier's full procurement cycle from receipt of an order to completion of all liabilities including provision of spares. Any problem areas discovered are followed up later.

10.14 PRE-QUALIFICATION OF SUPPLIERS

One of the first actions to be taken at the commencement of a project is the production of a supplier's list. The supplier's list may be produced from existing company lists including owner nominations and preferences, or from a project specific list. Where a company list exists, it is frequently the case that some form of selection process will be necessary to reduce the number of potential bidders by a pre-qualification questionnaire. The pre-qualification questionnaire is a simple method of producing bid lists and will in most cases be the sole method used.

The pre-qualification questionnaire can either be of a general nature covering total project requirements, or be product specific; it is issued with the objective of producing a bid list(s) of companies sufficient to ensure adequate competition, and including those companies most capable of providing the goods at the most economical cost, while maintaining quality and schedule requirements. For high value or complex items, it may be necessary to supplement the pre-qualification process with interviews and/or visits to prospective suppliers.

Reference should be made to Chapter 9 covering contractor/subcontractor selection for the necessary requirements in formulating the tender list, evaluation, audit requirements, etc. where these become mandatory under the EU directives.

10.14.1 Evaluation

The pre-qualification questionnaire addresses general company details, finance, experience, references, capabilities, resources, QA, safety, etc., and each question can be scored or weighted as required. For evaluation of responses, either an objective or subjective method can be used. With an objective method each question is scored (normally out of 10) and a weighting given to reflect relative importance. Of the three principal areas to be assessed, points to be covered are as follows:

1. financial: last three years' accounts, bankers, parent company details;
2. technical: product range, previous projects, design and testing facilities;
3. resources: manufacturing details, location, types of personnel, equipment, etc.

10.15 TENDER LIST

As a result of the evaluation of the pre-qualification questionnaire, an agreed tender list will be derived. The number of tenderers required depends on complexity of items required and enquiry value. For most items, the tender list will vary between three and six companies.

10.16 ENQUIRIES

Enquiry content will vary depending on overall project requirements and the type and value of materials or equipment to be purchased. However, for a typical large size project, an enquiry will usually consist of the following documentation.

10.16.1 Technical

- Requisition: items and quantities required;
- data requirements: data required, submission periods, quantities, presentation method, etc.;
- specifications;
- packing requirements;
- QA requirements;
- inspection levels.

10.16.2 Commercial

- Covering letter: return date, delivery required, etc.;

- instructions to tenderers: submission method, copies required, details to be covered;
- form of tender: pricing details;
- terms and conditions.

Once issued, any queries from tenderers relating to the enquiry will be routed to the applicable buyer, who is responsible for their prompt answer. Any answers to queries that affect all tenderers will be issued as an enquiry bulletin and issued to all tenderers. Requests for extension to tender return dates will only be granted if programmes remain unaffected and where several tenderers are making similar requests.

Where enquiries have been issued on a 'sealed tender' basis then all such tenders will be opened at the same time and an initial record of prices made. The level of sealed tenders varies between companies but most tenders with a value more than £50 000 will be subject to this procedure. For low value enquiries, tenders can sometimes be accepted by fax to speed order placement.

Once opened, copies of tenders are distributed by the buyer to the various specialists for their technical analysis that continues in parallel with the commercial evaluation.

Where several companies have submitted tenders, it is usual to shortlist one or two companies at an early stage, to speed up the selection process. Once this has been accomplished by way of an initial analysis and agreement with the tender assessment team, the buyer will concentrate his analysis on the shortlisted tenders.

10.16.3 Analysis

On completion of the short list, the buyer and engineer will produce separate detailed analyses of the tenders. Any queries with relation to the tenders will be addressed to the tendering companies and replies will be included in the necessary analysis. Although the commercial and technical analyses are produced separately, a joint recommendation is made on completion, i.e. the best commercial offer that is technically acceptable. Apart from reviewing the initial capital cost the commercial analysis will also address the following items:

- price validity;
- delivery period;
- spares;
- capacity;
- financial stability;
- ancillary costs.

In addition to the above, any qualifications to terms and conditions, warranty period, terms of payment and the like will be considered and the effect on the purchase order evaluated.

Where necessary, meeting(s) will be held with tenderers to clarify both commercial and technical aspects of bids, so that a comprehensive recommendation to award can be made. The buyer will arrange and chair such meetings and will ensure that minutes of meetings are produced and agreed with the tenderers. Normally tenderers will be required to delete any of their standard supply conditions which may have been included in tenders.

10.16.4 Award

Once the analysis has been approved, the buyer will prepare a tender report and purchase order for approval and signature. The order will reflect the costings contained in the analysis and any technical agreements reached subsequent to enquiry issue. Orders are signed and issued once the necessary purchase requisition has been received from the engineer. If a pre-award meeting is necessary, this will again be arranged by the buyer and agreements reached reflected in the actual purchase order. Two copies of the order will be sent to the supplier, with one copy acting as the official order acknowledgement to be signed and returned to the buyer.

Any change of a commercial or technical nature to equipment or materials after an order has been placed, will be covered by the issue of an official amendment to order. The buyer will normally retain responsibility for the administration of the order through to order close-out.

10.17 EXPEDITING AND INSPECTION

10.17.1 Expediting

If suppliers and purchasers carried out all of their agreed obligations on time, there would be no need for the expediting function. However, this is often not the case and expediting is used to progress suppliers and keeps a project aware of potential delivery problems.

Expediting is undertaken to suit the programme requirements of a project and is concerned with not only physical delivery of equipment and materials, but also the flow of design data. For projects with extensive design requirements, the need for drawings and data from suppliers is of paramount importance to maintain overall project schedules. Therefore, before award of an order, the buyer will agree document submission dates with the supplier and these dates will form the basis of any required expediting effort.

The type of expediting to be used will vary from visits to the supplier's premises to regular contact by telephone or fax, or a combination of both. For visits, the expediter will produce a detailed report showing documentation status and order progress. Problem areas will be highlighted, remedial actions proposed (if necessary) and a forecast of delivery made. Document status will also be updated with latest promised dates for submission.

Expediting forecasts are fed back into the project programme so that problem areas can be immediately identified. Not all slippages in delivery dates cause problems, and areas that affect programmes are concentrated on. As a supplement to direct expediting, suppliers are often requested to submit regular (normally monthly) progress reports.

Frequency of expediting visits and order progress is reviewed on a regular basis so that its need and effectiveness can be measured and priorities reviewed. Where necessary, normal expediting visits can be supplemented by either management meetings or visits, to reinforce the expediting process.

The expediting function is also responsible for ensuring that equipment and materials are delivered in the correct quantity and to the right location. The expediter acts as the liaison point between the supplier and the delivery site, and forewarns sites what materials and equipment items are being delivered and when. Nothing moves from a supplier's works without the expediter's prior knowledge and permission.

10.17.2 Inspection

Inspection can be regarded as an industry assurance service in that its function is to ensure that what has been ordered is in fact what is supplied and delivered. In a perfect world such a service would not be required and items delivered would be to the correct specifications and quality; unfortunately this is not the case and inspection is required to 'police' the manufacture of purchased materials and equipment.

Inspectors will typically have a manufacturing or engineering background and particular practical and technical skills and qualifications. It may be necessary to employ third party inspectorates to undertake inspection on behalf of a purchaser when the item to be inspected is of a specialist nature, or due to the geographical location of the manufacturer.

Inspection establishes acceptability of materials and equipment through compliance with design, destructive and non-destructive testing and pressure, proof, load, burst, performance, functional and operational testing practices supported by relevant certifying statements.

Inspection is required not only to satisfy the requirements of the purchase order but also to ensure compliance with statutory regulations. The level of inspection to be applied depends on the nature of the requirement and on the complexity of the equipment and the criticality of the item to the project. Therefore, bulk materials are usually subject only to a final inspection, while a compressor package will be inspected at various stages throughout the manufacturing and testing process. To achieve this a manufacturing quality control plan is produced by the supplier for approval, which alerts the supplier to the level of inspection required during the life of the order and which contains witness, hold and surveillance points at which inspection is required before continuing with manufacture or delivery.

Inspection by the purchaser should not be allowed to erode the supplier's responsibility to ensure that what is manufactured is to the required quality and specification. Reports are produced after each inspection visit to highlight problem areas and to serve as a progress check on the order. On completion of manufacture, the equipment will be inspected against final approved documentation and if acceptable, an inspection release note will be issued allowing preparation for shipment. If problems exist, a 'punch-list' of items requiring rectification or completion will be produced by the inspector, and the supplier will be required to clear this list before reinspection.

As part of his activities, the inspector will check all appropriate certification and documentation and ensure that completed data dossiers comply with project requirements.

10.18 SHIPPING/TRANSPORT/INSURANCE

10.18.1 Shipping and transport

On most projects, suppliers are responsible for the delivery of equipment and materials to site. Once an inspection release note has been issued, the expediter will obtain details from the supplier of weights, sizes and quantities and will issue precise shipping instructions. Details will be advised to site to forewarn the site management of impending deliveries. While in transit, the supplier remains responsible for the equipment and arranges necessary insurances.

For large offshore projects or overseas projects, the services of a freight forwarding company are often used to control and execute the movement of equipment and materials. The use of a forwarder increases the control a project has on site deliveries, reduces the cost as a consequence of bulk shipments and allows priorities for movements to be set, so that urgent site needs can be rapidly addressed. At every stage in the delivery process, the project is aware of status and actual location of the equipment and materials. Advanced planning of abnormal loads takes place so that routes are identified and necessary permits obtained. It is usual for the forwarder to provide an in-house coordinator to carry out this work and liaise with the project team. Freight savings are identified by comparing the forwarder's price with that of the supplier's for deliveries.

Whichever method of delivery is used, the purchase order will clearly state delivery terms based on the internationally recognized Incoterms nomenclature. For deliveries by way of freight forwarder, orders will be placed on the basis of 'ex-works loaded onto purchaser's transport'.

10.18.2 Packing

Any special packing, packaging and labelling requirements will be covered in the purchase order and will take account of site conditions, storage, etc. For offshore projects, export packing is often specified so the equipment can be stored safely for several months. Unless packing requirements are specified by the purchaser the supplier's standard is usually accepted.

10.18.3 Customs and Excise

All imports into the UK from the EU are subject to VAT, while import duties may apply to goods from outside the EU. Where equipment or materials are designated for offshore projects, the purchaser can reclaim VAT paid and is exempt from import duties through inward processing relief (IPR). Because of the 'single market' within the EU, details of EU imports now have to be provided by the purchaser on a monthly basis to the customs authorities. Where a freight forwarder is employed, it is usual to include all customs records and liaison in their scope.

10.18.4 Despatch of goods

Once materials and equipment are released for despatch, the expediter will advise either the site direct or through a material controller, that items are available for delivery and he will obtain details of weight, size, etc. plus estimated time of arrival. On receipt at site, items will be checked and inspected to ensure that all items have been received, that they are in accordance with the stated requirements and that no damage has occurred in transit. Apart from checking the actual goods, paperwork will also be checked to ensure that the necessary certification, documentation and inspection release notes have been received.

If the items are acceptable, a report will be issued confirming that the material has been received (sometimes called a material received report (MRR) or goods received note (GRN)), which will allow any necessary payments to be made. If the items are incorrect or damaged, or if there are missing items, etc. then a report (sometimes called a damage discrepancy report or DDR) will be issued by the material controller to the buyer for action. The buyer will discuss any necessary action with the supplier and payments will be delayed or reduced until the matter has been cleared.

It is important in the case of damage being sustained, or goods being lost, that the applicable insurance procedures are immediately activated.

10.19 PROGRESS MEASUREMENT AND REPORTING

Measurement of procurement effort or progress is not an exact science and is accomplished in a variety of ways, none of which is totally satisfactory. The following methods are currently in use:

- weighting items in the procurement cycle such as enquiry issue, order placement, etc., and then allocating total agreed man-hours accordingly. Progress is then measured against these milestones for the total procurement effort;
- setting overall milestones for discreet activities such as dates for all enquiries issued, all major equipment delivered, etc. Although this provides defined targets for procurement effort, it makes no allowance for actual practicalities of achievement;
- monthly targets for achievement of activities such as order issue;
- assessing man-hours used on a monthly basis against agreed man-hour total.

For reports, the procurement team generates individual order reports (expediting plus inspection), overall status reports on orders placed plus pre-order stage and a project order register. Reports can be tailored to suit individual project requirements so that the complete procurement activity is managed.

BIBLIOGRAPHY

Baily, P.J.H. (1991) *Purchasing Systems and Records*, Gower.

Operational Maintenance

11

Replacing the fluid
catalitic cracker at
Pembroke Refinery,
Pembroke, Wales.

**Plant Owner Pembroke
Cracking Company.**

11.1 INTRODUCTION

11.1.1 Capital and revenue expenditure

When a plant is built, the costs for the design and construction are usually made available from a capital budget for investment in new assets. One feature of the design phase is the consideration that must be given to future running and maintenance aspects. The owner's loss in production, and therefore earnings, during a shutdown can be enormous, and maximization of reliability and reduction of maintenance effort must be incorporated into the plant design. This is a subject in itself and is not covered here.

The plant will be built by a construction team, usually through a contractor, who will then hand over the plant to the commissioning team. Commissioning means starting the plant up and running it under normal operating conditions to check that the construction and the process conform to the specification and design. Commissioning work is very specialized and complex and will not be dealt with in this document. The costs of the commissioning phase form part of the capital budget.

Once the plant is producing and meeting the process specification in quantity and quality, the plant is handed over to the organization which will operate and maintain it on a day-to-day basis. This organization is normally the owner, but may be a separate company which operates the plant without owning it. From this point onwards the costs are not capital costs, but are 'operating' or 'revenue' costs.

A budget is normally set aside each year to cover the running costs of the plant, part of which is the operating cost, i.e. salaries for the operators, electricity, oil, gas and water to keep the plant running, and the feedstock which is the raw materials from which the product is made. The other part of the operating budget, and the part with which this chapter is mostly concerned, is the maintenance budget. This is the money required for repairs to, and servicing of, the plant.

Maintenance can be further split into two subsections; the first is where the repairs and servicing can be carried out while the plant is still running and producing; the second section is where the plant must be stopped and made safe before the repairs and servicing can be performed. There are many names for each type of maintenance but we will distinguish them here by calling the first 'routine maintenance' and the second 'shutdowns'.

In the past, the generally accepted principle for maintenance, either routine or shutdowns, was to carry out preventive maintenance, the theory being that if equipment was kept in the best working order then the incidence of breakdown would be reduced. The current thinking, which is reported to have started in the USA, is that some aspects of preventive maintenance make it a very expensive and unreliable method of risk prevention, and that it is more cost effective to repair equipment only when a breakdown occurs. ('If it ain't broke, don't fix it,' is the US catch phrase for this concept.)

11.1.2 Routine maintenance

This is sometimes referred to as 'running maintenance', 'on the run maintenance' and 'day-to-day maintenance' among many other names. Routine maintenance covers all aspects of work required to ensure that the plant runs smoothly and efficiently and ranges from grass cutting of landscaped areas to window cleaning, from light bulb changing to major electrical works and from vehicle maintenance to complex compressor/turbine alignment. There are various reasons why maintenance needs to be carried out, and these will be covered in greater detail later in this chapter, but in essence may include the following:

- items may cease to work;
- items may work inefficiently;
- items may work unsafely;
- items may work unreliably;
- items may be costing too much to operate due to a fault in design or performance; and
- items may be creating unacceptable environmental conditions such as odour, smoke, greenhouse gases, pollutant exhaust gases, etc.

The decision as to whether maintenance work can be carried out while the plant is running would normally have taken place when the plant was first designed and would depend upon such factors as the cost of duplication, the criticality of the item and the type of process, e.g. continuous running or batch processing. Examples of continuous running would be refineries or coal-fired power stations where starting and stopping are very expensive.

Batch processing is frequently found within the food industry (e.g. bread baking), in the pharmaceutical industry or iron foundries. With batch processing, maintenance can generally be performed between batches. Normal maintenance must still be carried out, however, to ensure that the equipment is functional when required.

In the case of continuous processes, there is no opportunity to carry out maintenance between batches and therefore the design of the plant must consider all practical options to reduce the risk of one item disrupting the entire production.

In a refinery, for example, it may be cost-effective to construct a section with two pumps where only one is actually needed and one is spare. These pumps are usually referred to as 'duty' and 'standby' respectively. This option allows one pump to be maintained while the other one works. Although this may be practical with certain smaller items, it would be prohibitively expensive to design a plant with a spare high pressure, medium pressure and low pressure turbine/compressor configuration on the off chance that one might break down during operation. Hence, risk and cost are major factors in design criteria with regard to maintenance considerations.

Standardization of equipment is another essential, enabling an owner to keep only one replacement for each type of unit rather than one for each item of plant, effecting savings in revenue and storage costs.

11.1.3 Shutdowns

Again there are several alternative names given for shutdowns. Common names include shutdown, outage, overhaul, turnaround, stop or other epithets. In this case we will consider all to mean the same, although some organizations and indeed individuals have different meanings for each. Some consider a shutdown to be only when the plant is stopped finally ready for dismantling, while others consider a shutdown to be when the plant stops unexpectedly. It again depends very much upon the individual.

Planned shutdowns are necessary where maintenance work cannot be done as routine maintenance while the plant is running. Examples of this are furnaces or boilers in power stations and distillation or fractionation towers in refineries. In many cases shutdowns are not required by the owner or even as a requirement of the process, but are actually to conform with statute. There are different laws governing various sectors of industry, but generally these laws have evolved to ensure that the operators and/or the public are safeguarded from failure of any equipment and subsequent leakages or explosions. One of the main operations carried out during shutdowns is therefore mandatory inspection to comply with these laws.

Many processes cause fouling of equipment and this must be cleaned regardless of legal requirements. Another common reason for shutdowns is to incorporate improvements in process or debottlenecking of plant. As the strength of a chain is dependent on its weakest link, so the throughput of a plant is governed by the most constrained item of equipment. This is known as the bottleneck. Debottlenecking is the removal or reduction of these constraints (this cannot be repeated *ad infinitum* as the plant is only designed to a certain capacity and continual upgrading becomes less cost-effective).

The timing of a shutdown is normally governed by statute and market forces. The law may possibly be stretched by formally applying for a delay, but market forces may not. It is usual to plan well in advance when the next shutdown will take place. All jobs which cannot be completed with the plant running will then be listed as possible jobs for inclusion in the shutdown works. Work on the shutdown will be limited as far as possible, to minimize loss of production and profit. The shortest shutdown period is the optimum target.

11.2 PLANNING

11.2.1 Scheduled routine maintenance

Scheduled routine maintenance can be related to the regular servicing of a car. It is the same concept in any process plant; the components may be different, but the principle remains the same. Pumps may require new seals, valves require new packing or seats, or other consumable items may need replacing. This work can often be carried out either while the plant is running, or during a plant shutdown. In each case, it is vital that the maintenance is properly planned.

In order to carry out certain maintenance while the plant is running, there may have to be a pre-installed back-up component, for example, a secondary standby pump which can be run while the main pump is dismantled. This work must be carefully planned to ensure that maintenance is carried out before the components reach such a state that the equipment is damaged or fails completely. The planners must devise a schedule, calculating the optimum running time of each component which requires to be serviced, and distributing the tasks to avoid peaks and troughs in workload.

The means of planning is normally to ascertain the amount of work in man-hours to help determine the minimum number of personnel required to keep the plant functional. It is no use producing a schedule which requires 140 men every third week of the year and only 15 men in the intervening weeks. This would be an exceptionally expensive way to employ labour, whether it is in house or supplied via a contractor.

In the event of a scheduled shutdown, even more emphasis is required in planning. Many plants either feed, or are fed by, other plants, and these other plants must be considered when deciding the date and duration of a shutdown. To give an example, oil installations in the North Sea are continuously producing oil and gas. The oil may be constantly required for refining and is the sole feedstock for a refinery. The gas produced at the same time may be separated and piped to an ethylene plant which extracts the ethylene and sends it to a polyethylene plant which makes polythene beads, which are in turn shipped to manufacturers to make gas and water pipes or even polythene bags for supermarkets. If it is decided to shut down the ethylene plant, it must be decided what happens to the vast quantities of gas still being produced from the North Sea, how to feed the polyethylene plant and how to maintain production of pipes. An obvious answer would be to have large storage facilities for each stage, but this is an uneconomic solution given the low prices, the competition for each product, and the immense storage areas which would be required.

Planners must therefore liaise very closely with the 'business' units of each plant. These units look after the owner's commercial interests, such as deciding from where the feedstock and back-up supplies may be bought, and what their marketing commitments are for their own products. The whole chain must be consulted before a date can be arranged. Usually the business units dictate to the planners when each unit can be shut down.

The next consideration is the duration of the shutdown. Again, this is frequently set by the business units. They may be able to buy feedstock at low prices for a certain period from alternative sources, or they may be given a period of dispensation during which their customers will find alternative suppliers. Contract commitments and market forces are usually firmly set regarding these periods. It is therefore vital that all work on a shutdown, or the portion of work which can only be carried out during a shutdown, is planned to be carried out within the given time.

The work required, such as to vessel internals, can usually be assessed based on historical information, but some rearrangements to the programme may be necessary should the extent of work found, on opening up, to be more or less than anticipated.

To keep the shutdown period to a minimum each task must be reviewed and consideration given to methods of working which will maximize the preliminary work, prefabrication and the like which can be undertaken before the plant is shut down. Using alternative methods of work which reduce the workload in the shutdown period may increase the total cost, and this must be balanced against the reduced time that the plant is non-productive.

The planners, therefore, have to look at every task to be carried out and calculate, using man-hour norms, how long each task will take. They must assess how many men can work simultaneously on each task, thereby working out the duration of those tasks. Many plants are very complex and congested and it is not always possible to use the theoretical optimum manning levels for reasons of safe working and accessibility. There are only so many men that can be used efficiently on each task.

The planners must look at all the tasks, the number of men who can work on each, the duration of each, and how many tasks can run simultaneously. Certain tasks depend on the previous one being started or completed before work can start. The optimum will eventually be fixed by using various planning methods which it is not appropriate to describe here; suffice it to say that planning is a science on its own. There will be one sequence of tasks which cannot be reduced in duration and is dependent on other tasks finishing. This is called the critical path. The critical path cannot be shortened or reduced and this determines the length of the actual shutdown. The duration given by the business group may be sufficient to carry out the work under normal shift working or may even require seven day working, 24 hours per day. Where the critical path cannot be fitted into the time scale, then negotiations are required to realistically extend the time allowed.

With good planning and critical path analysis combined with the time allowed by the business unit, the planners in conjunction with specialist engineers can work out the total number of men required on the site and the shift pattern they are required to work. This information is fundamental to contractors for resourcing the task, and to the cost control personnel for fixing budgets.

11.2.2 Scheduled particular maintenance

Scheduled particular maintenance can be described as a major element of work which was not envisaged in the long-term running of a plant. The work is essential and can be scheduled to be carried out alongside the routine maintenance while the plant is still running. Particular maintenance always puts a strain on resources, and effective scheduling of the work to fit in with other maintenance is critical. An example of particular maintenance would be where moisture has penetrated between insulation materials and pipework. Due to capillary action, this water will spread over many hundreds of metres, causing corrosion to the pipework, which will be imperceptible to the naked eye as the outer insulation will be unaffected and the point of ingress may be concealed (underlagging corrosion).

The maintenance involved here will be the stripping of all insulation, determining whether the pipework can be blasted and repainted or has to be replaced, and the reinsulation of the piping (possibly to a higher specification) to prevent recurrence. The stripping and reinsulation could be performed while the plant is still running, provided the insulation is not critical to the process.

The planning of this particular maintenance must take into account the insulation contractors' available resources, the contractors' expected workload, whether a discrete contract should be awarded and work by other contractors in the same area. Should an underestimation be made regarding other insulation work being concurrently undertaken by the contractor, then overruns of costs and time, and even loss of production, may occur.

11.2.3 Exceptional maintenance

It is often difficult to distinguish between exceptional maintenance and particular maintenance and many owners use only one expression to cover both categories. The overlap often makes the distinction impossible even when both descriptions are used. Exceptional maintenance may be described as maintenance which was not anticipated in the long-term running of the plant (the same as particular maintenance), but is not scheduled and cannot be carried out over a long period. Exceptional maintenance may be carried out solely to prevent breakdowns. A description of breakdown maintenance is given in the following section.

An example of exceptional maintenance would be where the feedstock of a plant changes, causing a pump to become clogged with sludge as the new conditions are outside the pump's design criteria. The pump would become very inefficient and may affect the output of the plant.

Planning the cleaning or replacement of this pump must not affect the routine maintenance of the rest of the plant. Again, the availability and skills of the contractor's resources must be considered, and also the concurrent maintenance requirements must be predicted.

11.2.4 Breakdown maintenance

As the name suggests, this is the maintenance required when an item of equipment fails. If the equipment has a standby, then production will not be affected and the repair time is not so critical; but if there is no standby the plant will cease production. The planning of this work not only involves the repair and the possible restart of the plant, but also the prioritizing and rescheduling of all other work which is deferred due to the critical nature of the breakdown.

11.2.5 Constraints

Process plants are generally very congested with pipework, cabling and equipment. The efficiency with which work can be carried out is very much reduced in comparison with the construction of a new plant. In addition, on some plants, the nature of the chemicals and gases contained in the process are extremely hazardous to health, and exposure to them is only permitted for a limited period even with protective clothing and breathing apparatus.

The nature of work on live plants, such as working within vessels and towers which are very cramped, and the requirement to have another person standing by outside for safety reasons, and for working in protective suits with breathing apparatus, are all components of very inefficient working. Other constraints include evacuation of the area prior to radiography of welds, restrictions on flame cutting of pipework or steel, the use of non-sparking tools and welding only in appointed safe areas, all of which must be evaluated when planning the work.

The above situations, to varying degrees, are all present in process plants and it is the planner's job to calculate how each situation affects the man-hour norm and productivity. Each plant or part of a plant will have a different factor to apply to norms to develop an accurate overall plan. An accurate estimate of price is extremely difficult.

11.2.6 Coordination

As with a new project, a programme must be developed for all the known maintenance work. Should a breakdown occur, or exceptional maintenance be required, this programme may have to be temporarily abandoned for the more critical work to be performed. This often requires the involvement of more than one contractor. For one mechanical job to be executed there is frequently a requirement for scaffolders, insulators, electrical contractors, inspectors, painters and non-destructive testers. The operation of these contractors must be strictly coordinated to minimize the time taken by each and the interference with others. Even when a job is critical, the time taken by the planner to produce a programme is more than saved by the increased efficiency of the coordinated working.

The provision of materials, spares and components all must be planned, pre-ordered and expedited to ensure the work is not delayed.

All contracts for maintenance work should address the amount of coordination required between contractors and with the owner and should make allowance for unforeseen incidents which necessitate flexible working.

11.3 ESTIMATING AND BUDGETING

Estimating generally is covered elsewhere, and this chapter aims only to refer to estimating and budgeting particular to maintenance.

11.3.1 Estimate

The estimate is a prediction of the final cost of work. In its preparation it requires a blend of experience, historical data and a forecast of commercial conditions applied to a technical understanding of the work.

The accuracy of the estimate will depend upon the level of technical definition, the quality of the estimating data and the assessment of the risks inherent in the maintenance work including realistic allowances for unknowns.

11.3.2 Sources of cost information

Historical data

Records of past maintenance work and purchases usually contain extensive and useful cost information. If such information is systematically analysed, recorded and filed together within a suitable retrieval system, this forms a good basis for costing future work.

When utilizing historical data as a basis of an estimate, it is important to consider the following points, both in relation to the current project and to the project from which the historical data were extracted:

- procurement strategy;
- reliability of information;
- productivity factor;
- similarity of the planned work (local conditions and time scale);
- local conditions;
- time scale;
- condition of the plant.

Unit costs/norms databases

Estimating data for operational maintenance has generally been difficult, with each owner employing differing methods of measurement, forms of contract and specifications. Unlike the building industry there is little published information available with regard to pricing. However, published labour norms exist for the majority of disciplines in the construction of process plants, and these may be adapted for use for maintenance work.

The major publications are produced by:

- Page and Nation;
- Oil and Chemical Plant Constructors Association (OCPCA).

When using such information it is important to consider all items included under 'historical data' and the following:

- the costs/norms should be for maintenance work and not new construction;
- the amount of work, e.g. a small quantity of work will be more 'costly' per unit than a large quantity of work.

The above norms are generally factorized in accordance with the working environment and labour productivity.

11.3.3 Management costs

On new projects the management is frequently dedicated to the work and thus all costs are charged to the project. With maintenance there may not be any staff dedicated to one particular area or cost centre. When estimating the overall costs of maintenance it is necessary to include the costs of management in the correct cost centres, or their actual costs may cause cost overruns when finally allocated.

Owners do not always apportion their own managers/staff costs to maintenance, as these are recognized as an overhead cost to the business. In the case where owner's staff are dedicated to the shutdown, other personnel are often contracted in to fill their normal roles. The costs of these contracted personnel must be included in the overall shutdown estimate although they do not appear to be a direct cost. Similarly, and more obviously, if a managing contractor is appointed, the costs must be allocated to the job being managed.

11.3.4 Numbers of contractors

The number of contractors employed on routine maintenance, or a shutdown, has a direct effect on the costs thereof. For each contractor there is a level of management within the costs. If several contracts can be added together as one large contract, there is only a single management component to be included, albeit larger than each individual one, but smaller than the sum. Conversely, if the contract becomes too large, the required contractor's management resources may necessitate the use of a large multi-functional contractor with the attendant large overheads and this could be inappropriate for the types and amount of work required.

The use of many contractors involves greater coordination by the owner to prevent disruption.

These considerations must be taken into account when preparing maintenance estimates, budgets and contract philosophy.

11.3.5 Reliability of costs

Due to the various factors and working conditions affecting maintenance, any historical costs or norms must be carefully evaluated to ensure that they are appropriate. Simply taking a norm for replacing seals in a pump would be incorrect if the man-hour norm was for a pump in lubricating oil service and the same man-hours were being applied to work on a pump in unfiltered crude oil service. Furthermore, all prices must be checked to determine the build-up of that cost.

Management costs may have been excluded from individual equipment costs on previous jobs but are to be included per item of equipment in the estimate. It is also important to ensure that the prices and costs in any historical data are actual costs expended and not only forecast costs which were actually overrun. Prices of spares and replacements should be checked with the manufacturer to ensure that large price changes have not taken place since the previous expenditure. The existence of spares price agreements with manufacturers should also be investigated.

11.3.6 Category of maintenance

For various reasons, many owners wish to know in what category of maintenance costs are being expended. Estimates will have to be divided into each of these categories per item of work. This categorization is in addition to the cost centres.

Several of the more common categories include:

- reliability: maintenance is required because the item of equipment is unreliable in operation;
- statutory: equipment must be opened up and inspected in accordance with statutory requirements, e.g. pressure vessels;
- improvement: the equipment requires modification to improve the product or process;
- inspection: inspection department have requested that the equipment is opened to check corrosion, erosion, etc., usually as a result of problems encountered in a similar environment elsewhere;
- safety: the equipment requires inspection or modification if there is a possibility of fracture, leakage, etc;
- performance: the equipment requires repair or modification to improve its performance or output;
- environment: the equipment requires some work to reduce pollutants being introduced to the environment.

Some of the above will make the plant more efficient and will pay for themselves in time, while others will have no financial benefit and must be written off as a direct loss but will have indirect benefits, e.g. safety, environmental, etc.

11.4 CONTRACTING STRATEGIES

11.4.1 Introduction

One of the stages in engaging a contractor to undertake maintenance work is the preparation of the tender documents. It is at this stage that the apparently straightforward matter of choosing the type of contract can influence the economic success of the project. A contract strategy must be developed which will reflect the type and amount of work, the complexity and reliability of the plant and the time available for procurement and performance of the work.

11.4.2 Facilities management

Facilities management is a method of maintenance where all the responsibility for the required availability of the plant is outsourced to a contractor. Before tendering the contractor is given the specification of the plant, any available historical data of the previous maintenance and the operating requirements. The contractor then submits a price for maintaining the plant to a certain standard for a certain period. Payment terms may be lump sum or resource based. This type of contract is rarely used on larger process plants, but is becoming more common on manufacturing and production-line plants.

10.4.3 Routine maintenance

By its very nature, routine maintenance work will cover planned and unplanned activities, work which is both quantifiable and unquantifiable and work which may have to be undertaken outside normal working hours. In all cases the principal requirements of any contract should be 'flexibility'. Due to the unpredictability of breakdowns, even with the best planning, peaks and troughs in workloads are certain to occur and any contracting strategy should address and allow for these variations.

11.4.4 Shutdowns

To prevent unnecessary loss of production it is essential that shutdown time is kept to a minimum with as much preparatory work as possible being performed before the plant stoppage. Any shutdown contract should be flexible and require the contractor to supply additional resources for any extra work arising as vessels are opened and inspected. The contractor would still be required to finish the work without an extension of time.

Planned/unplanned

Shutdowns are either those planned for maintenance or the unplanned stoppage of plant resulting from breakdown.

Shutdowns resulting from planned activities can be accommodated into a form of contract where control can be applied through a schedule of rates and/or reimbursable costs. Lump sum contracts are rarely successful due to the probability of emergent work on opening equipment. Unplanned shutdowns may require a totally flexible approach where a reimbursable contract may be the only option.

Term contractors

Existing term contractors can be used for work on shutdowns. However, there may be a danger of duplication of resources, i.e. payment by manhours on one contract while the same is being paid on a schedule of rates elsewhere. In addition, term contractors may not have the resources or skills to carry out shutdown work.

Partnering

Reference should be made to partnering as described elsewhere in this book. This approach is not recommended for shutdowns due to their relative short-term nature and the long-term requirements of partnering, unless the shutdown forms part of a larger maintenance contract.

In-house

An in-house maintenance department will rarely have the capacity to handle the whole shutdown in the time required or it will be executed at the expense of other maintenance. It is more usual that in-house maintenance handles discrete areas of the shutdown where specialist knowledge is required, e.g. instrumentation, turbines/compressors.

Management contractor

The services of a management contractor are often incorporated into the management of the shutdown. The appointment should be as early as possible to enable proper planning and coordination of the work. The management contractor provides the main management team and can also be the executing contractor in some cases.

11.4.5 Methods of payment

Generally the payment for maintenance or shutdown contracts falls within one of the following three main categories:

- reimbursable (basic cost plus management fee);
- lump sum;
- schedule of rates.

There are many derivatives of these payment methods, and contracts can contain more than one category. Reference should be made to Chapter 7 on contract conditions for more detail of these contract types.

11.4.6 Disaster recovery

This is the approach and the activities required to bring a shutdown or maintenance work, either back on programme or within budget or both.

The principal requirements will be to identify the problem(s), prioritize the schedule, allocate resources, put in place a definitive reporting and monitoring system and continually review progress against critical activities.

Every contract placed for shutdowns or routine maintenance should contain provisions to allow for this corrective action without determining the contract.

11.5 CONTROL OF COSTS

11.5.1 Introduction

The fundamentals of cost control on maintenance projects are:

- cost monitoring: a procedure capable of providing reliable information in due time;
- cost control: the decision making in response to the information provided by the cost monitoring procedure.

Cost information is required for a maintenance project on a continuous basis:

- daily and/or
- weekly and/or
- monthly.

Without this information, management has lost a major tool.

The main components of cost control are:

- good estimating procedure;
- effective cost code or cost centre system;
- workable variation system;
- efficient 'actual cost' reporting system;
- practical and effective procedure for predicting trends and final costs.

It is the last component that reveals where things are going wrong. Without this component the system merely monitors. To get to this fifth stage one must progress through the first four stages and then apply judgements. Where necessary, decisions and changes are made to correct any problems to bring the job back on to time and budget.

11.5.2 Work orders

On term maintenance contracts and occasionally on shutdown contracts there is no definitive scope of work contained within the contract documents. The contractor is required to perform work as and when instructed by the engineer.

To control costs on these types of contract, there has to be a written instruction to the contractor which clearly describes the work, gives the starting date and time, identifies the method of reimbursement and to where the costs must be allocated. This information is necessary for budgetary control on particular items of equipment, processes or geographical areas of the plant, depending on the cost reporting requirements. For historical purposes it is also essential that the exact workscope, costs and details are recorded, so that future estimates will be accurate.

The most efficient method of instructing the contractor is by a work order system. There are several names used for a work order, such as site instruction, contractor's instruction, authority to perform work, etc.

The responsibility for the raising and issuing of work orders is that of the engineer/supervisor responsible for the work.

11.5.3 Variations

The control estimate should contain a contingency sum which will include for some emergent work which is likely to be identified on opening and inspecting the equipment. In the maintenance sector, however, unforeseen items of work and emergency repairs frequently occur. As this work is not within the predicted scope there may be no provision for it in the budget.

The project manager must decide if the cost of these scope changes can be contained in the contingency sum. If so, funds must be transferred out of the contingency to the relevant cost centre.

If the contingency sum is insufficient to accommodate the scope changes, the project manager must either apply for a supplementary sanction or show a cost overrun. It is essential for a change procedure to be implemented in order to list and record any variations which may occur. To control the maintenance costs, all changes are recorded and their effects properly evaluated.

The programme and budget can then be updated as the changes are authorized. Contractors naturally base great importance on recognizing changes which will affect cost and programme, and will cause interface problems with other disciplines.

Variations can often result in disruption of the contractor's work with consequent potential for claims for additional costs and delays.

11.5.4 Measurement

Measurement is necessary to determine the true work scope against which costs are expended on all types of contract except lump sum. There is a school of thought, however, which advocates measurement of resources used and work performed on lump sum work to determine value for money and to assist in the accuracy of future estimates. But contractors are reluctant to expend effort in preparing these figures. Where these figures are required, it should be stated as an obligation of the contractor in the tender documents.

Measurement of maintenance work generally must be done as soon as possible as, unlike new construction projects, the work is rarely shown on drawings and many of the materials are reused rather than new. It thus becomes difficult to identify work in retrospect. Similarly, if work is being carried out on a reimbursable basis, it is necessary to monitor and report resources per item, and this cannot always be done in retrospect if records are incomplete. Some of the mechanical or specialized work may be performed inside vessels or towers, and it is usually beneficial to close up this equipment as soon as the work is complete, which makes subsequent measurement impossible.

Unless measurement is accurately performed and recorded the reliability of cost information is impaired. This in turn reduces the accuracy of the cost reports and future estimates.

11.5.5 Approval of work

Payment of any item of work is based on the work being accepted by the owner. During maintenance work, acceptance can sometimes become a grey area. If a contractor bids a lump sum for opening a vessel and cleaning for inspection, the degree of cleanliness must be specified or a disagreement could arise. The contractor will wish additional payment if he is asked to clean further, while inspection personnel will feel that the cleanliness they require is covered within the original lump sum.

On reimbursable contracts, attention must be paid to rework in instances where the original work is rejected. If a contractor fits bolts of the wrong material and then has to refit the correct bolts, the time for removing and refitting should not be payable, unless due to the owner's error.

11.5.6 Star rates versus dayworks

Contractors may tender very competitive rates in the schedule of rates in order to win work on a long-term maintenance contract. To minimize his risk, the contractor may then request the use of reimbursable rates on variations rather than the unit rates. Conversely the owner may have failed to define adequately some elements of the work, but will nevertheless endeavour to apply the unit rates in order to minimize his own risk.

Where an estimate is based on prevailing unit rates, cost control could suffer greatly should most of the work be paid on a reimbursable basis. For this reason it is essential that a lot of effort is expended in the composition of the schedule of rates to ensure that the items are comprehensive and accurately describe the type of work to be performed.

11.5.7 Disruption of contractor's work and potential for claims

Due to the changeable character of maintenance, uninterrupted work is not always possible if other breakdowns occur, either in the vicinity of the work, or requiring the use of the contractor's resources elsewhere. Disruption on a live plant could be caused by the presence of hazardous substances, toxic gases or other safety restraints. All these interruptions mean extra costs to the contractor which he will pass on to the owner, either in the form of extra work or as a claim for extra costs and delays.

Where the delays exceed the expectations included in the estimate, there could be a requirement for contract variations to be raised, although no extra work or changes to scope may have occurred. One of the difficulties in accurately forecasting final costs is assessing the likelihood of contractors submitting claims for such delays and disruptions at the end of the job, and the settlement value of these claims. Owner's representatives should carefully monitor all delays and disruptions to help assess and agree any such subsequent claims for additional costs.

11.5.8 Self-policing by contractor

Where practicable, some responsibility should be placed on the contractor for assisting in the cost control of the maintenance. Obligations and incentives should be included in the contract document to this effect. The greatest incentive for this control is the gain sharing approach where both contractor and owner benefit from more efficient working by the contractor.

11.5.9 Predicting trends and final costs

Predicting trends and final costs is a very important activity, as control depends upon the prediction of final costs for each item in the control estimate. Every judgement and forecast should be carefully considered by the cost engineer in consultation with the project planner.

11.6 CONTRACTS/FINANCIAL AUDITS

On completion of a maintenance job which is liable to be repeated sometime in the future, there is benefit in preparing a post contract review which should cover the successes and failures of the project.

11.6.1 Contract documents

A complete review of the suitability of all parts of these documents should be undertaken. The final work scope should be compared with that included in the documents and a true assessment should be made as to what was omitted from the original which should have been anticipated. For historical purposes there cannot be too much detail in the final work scope.

Any contractual problem which arose during the life of the contract should be noted and preventive measures included in future contracts. If a new contract is not needed for some time, then notes should be attached to the finished contract for reference. Description of priced items should be amended where disputes arose between contractor and owner.

11.6.2 Invoices and payment

The timely submission of invoices and their payment should be audited to ensure that sufficient back-up information was supplied and that enough time was allowed for payment. The apparent late payment of invoices can be the cause of the greatest discord between owner and contractor, although some blame can generally be placed on each party. The owner will normally reject invoices if the following problems are experienced:

- the invoice does not match the certified valuation;
- the invoice does not contain the correct contractual information, e.g. contract title, contract number, etc., or
- the amount is not correctly allocated to cost centres.

11.6.3 Price of non-conformance

In all aspects of maintenance, the price of non-conformance is considerable to the contractor and to the owner both in direct and indirect costs.

Non-conformance in quality during maintenance has a direct cost to the contractor in rectification costs and usually some consequential costs to the owner in delays, reinspection, etc. Non-conformance commercially, e.g. incorrect invoices, late payment of invoices, overdue final accounts and the like, create cash flow problems for the contractor and subcontractors leading to finance charges and loss of accrued interest.

Some of the other items on non-conformance include:

- late design information;
- design errors;
- emergencies (gas alarms, fires, etc.);
- late access to items of equipment;
- late delivery of materials;
- equipment in worse condition than anticipated;
- drawings of existing plant not up to date;
- inspection personnel not available when required.

All of these can cause serious delays which may result in disruption to the contractor's and the owner's programmes and necessitate disaster recovery action.

In an effort to reduce non-conformance to a minimum, techniques such as total quality management are utilized by owners and contractors to highlight, record and rectify areas requiring rework and modification.

Where non-conformance has been identified on a job, the cost, where possible, should be separated. Future contracts should address the problem if the cause lay with the contractor, and should include financial disincentives. The owner's organization should review its procedures if the problem arose in house.

The actual requirements should also be assessed as conformance may not be realistically achievable.

11.6.4 Definition of work performed

As stated earlier within this chapter, the detailed definition of work carried out is the single most important feature required for historical data. Many different jobs may be carried out on a single item of equipment and unless the exact details of each job and the number of jobs is recorded, incorrect assumptions as to similarity of future work compared with previous work will be made. The accuracy of estimates will be directly affected by these assumptions.

11.6.5 Final accounts

Final accounts represent the financial detail to be associated with the workscope details described above. Precise costs must be attributed to job details to ensure cost reporting and future estimate accuracy.

The final account should, where possible, be structured in such a way as to enable an analysis for statistical purposes.

11.6.6 Historical data

In many cases, cost information gets buried with the maintenance records, discouraging any search for cost data on a particular piece of equipment. It is therefore important to store all cost data and maintain a comprehensive database of historic costs, with the exact details of work performed, spares fitted, etc.

In maintenance contracts in particular, the working environment can change considerably due to the following:

- permit conditions;
- access;
- live plant conditions;
- nature of product;
- materials used;
- location.

For the above reasons historical information must be recorded, analysed, archived and utilized, giving due cognizance to all factors affecting the cost and more importantly the man-hour duration of each event.

Provided there is a standard format used for the archiving of job details, each job recorded will add to the overall accuracy of the data. The archiving format must be usable and data easily extracted for manipulation/factorization.

11.6.7 Update unit costs/norms database

Unit costs/norms for measured work are usually built up and maintained by each owner to cover all work over many years. It is essential that the collected data are constantly reviewed and updated in line with changes in working practices, use of plant, inflation, productivity, safety regulations and the like. Unit costs can be updated primarily as follows:

- use of published indices, e.g. the Association of Cost Engineers;
- review of rates from tenders received;
- review of final accounts.

The updating of man-hour norms is more difficult, and only by recording the durations of each construction task can these norms be revised.

Updating of statistical norms such as overall man-hours/metre of pipework, man-hours per tonne of steelwork and the like can be carried out fairly easily by analysing the final account for every contract awarded.

Nowadays, many owners are submitting information to organizations which compare costs, on a confidential basis, with their competitors and publish statistical charts and analyses.

11.6.8 Value for money

With the ever increasing drive to reduce costs, owners have been striving to get better value for money. Design and materials specifications are constantly being reviewed to ensure that costs match, but do not necessarily greatly exceed, engineering requirements.

Value management has now become a recognized activity, with more and more owners utilizing the services of value engineers, who assess the standards required and the work methods used. The attitude of 'We've always done it that way', is being questioned more now than ever. Preventive maintenance may become a thing of the past.

Health and Safety and the Environment

Blood Products
Laboratory, Elstree,
England.

**Laboratory Owner Central
Blood Laboratories.**

12.1 INTRODUCTION

There are currently between 50 and 60 Acts of Parliament and about 400 sets of regulations that are relevant to health and safety at work.

This chapter provides guidelines as to what safety and environmental legislation is applicable to the construction industry and gives guidance on what the responsibilities and entitlements are for all parties of a project.

The meeting of both safety and environmental legislation inevitably adds to the administration and construction costs of a project. These additional costs should be assessed and included in the project estimate.

12.2 HISTORICAL SUMMARY OF LEGISLATION

Legislation comprises Acts of Parliament and legislation made by subordinate bodies which have been given authority by Act of Parliament.

For more than a century health and safety legislation for persons working in the UK had developed a piece at a time in response to disasters and the high fatality rate in the construction industry, each one often covering a particular class of person and not in a consistent manner. Separate legislation with variations in details and in methods of enforcement would apply to a specific requirement when undertaken in a factory, as opposed to an office, a mine or a quarry.

Between 1875 and 1937 there were attempts to unify the increasing but fragmented legislation, culminating in the Factories Act 1937. However, regulations made under previous legislation continued in force causing confusion and misunderstanding.

The Factories Act 1961 repealed the shortcomings of its 1937 predecessor which was recognized as having a tendency to look to the protection of plant and equipment as a way of preventing injuries to workers. The Act of 1961 put more emphasis on the safety of visitors, contractors, neighbours and other third parties.

By 1970 many organizations were questioning whether the existing legislation was either sufficient or effective in providing proper protection for people at work. The pressure on government as a consequence resulted in the setting up of a review committee under the chairmanship of Lord Robens. After studying the problem in depth the committee reported in 1972, making many wide-ranging recommendations.

The essence of the Robens Report recommendations was to:

- replace the mass of existing safety legislation with one Act applying generally to all persons at work;
- replace the mass of detail with a few simple and easily assimilated precepts of general application;
- change methods of enforcement so that prosecution was not always the first resort;
- ensure that occupational safety should also protect visitors and the public;

- place more emphasis on safe systems of work rather than technical standards;
- actively involve workers in the procedures for accident prevention at their place of work.

The main recommendations of the Robens Committee were accepted by Parliament and were incorporated in the Health and Safety at Work Act 1974, and it is this Act which is the principal piece of legislation governing health and safety at work and maintenance of safety records today.

12.3 THE HEALTH AND SAFETY AT WORK ACT 1974

Most of the Health and Safety at Work Act 1974 came into force on 1 April 1975. It created a new approach to safety law by saying what should be done in the best interests of safety for people at work, setting goals to be met rather than being prescriptive, as previous legislation. The Act took up the theme of the Robens Report with a view to 'making safety everybody's business' and extended the responsibilities to visitors to the workplace, including contractors, subcontractors and any other person who might have cause to be in or near the workplace, including the safety of the general public.

Many people still think about the effects of accidents in terms of insurance and how much compensation might become due. The Health and Safety at Work Act is mainly concerned with those who create the risk controlling the risks. Those who do not are liable under the Act for prosecution, the penalties for not complying ranging from heavy fines to prison sentences.

The Act created a body – The Health and Safety Commission (HSC) – to become responsible for the development of health and safety legislation and its application at policy level. It also created the Health and Safety Executive (HSE) to be responsible for the implementation and administration of most matters concerning safety and the law.

12.3.1 Scope of the Act

The Health and Safety at Work Act is of overriding importance in almost all work situations today and its provisions should be considered before going on to the specific requirements of any other Acts and Regulations which may be applicable. The Act provides such wide protection that it not only requires employers to have regard to the health and safety of their own workforce but also requires protection for members of the general public who may be endangered in some way by the activities of the employer.

12.3.2 Duties

The Health and Safety at Work Act creates duties and responsibilities for everyone at work to set up and maintain a safe working environment. It specifies the responsibilities of employers and employees.

The Act also requires employers and self-employed persons to set up and maintain a safe environment in respect of other people, such as contractors, subcontractors, visitors to premises and the general public.

Compliance with the law and the creation of a safe working environment must begin with the development, by the employer, of systems of work that ensure safety for everybody and not just those they employ.

Managers and supervisors must understand and accept their roles and responsibilities for health and safety at work. Some are mentioned specifically in the Act and others in the duties that they have to carry out.

Every employee has a responsibility to cooperate with the employer in the maintenance of a safe working environment.

The duty of every employer

'It shall be the duty of every employer to ensure, so far as is reasonably practicable, the health, safety and welfare at work of all of his employees.' This is a quote from section 2 of the Health and Safety at Work Act. It does not make any technical requirements. It does not set any minimum standards of behaviour that can be measured. The setting of technical requirements and measurable standards of behaviour has to rely on what has happened in the past — what others have or have not done — and what can be done.

The minimum standard required is to do what is reasonably practicable to create and maintain a working environment that is safe in the light of what has happened in the past, what is happening now — and what someone at work is likely to do in the future.

To measure what is 'reasonably practicable' it is necessary to compare the degree of risks against the sacrifice (whether in terms of money, time or effort) to avert that risk. For example if the risk was insignificant in comparison to the 'sacrifice', then the duty holder will have proved that compliance was not reasonably practicable.

The responsibilities of the employer

An employer must so far as is reasonably practicable:

- provide and maintain plant and systems of work that are safe and without risks to health;
- have arrangements for ensuring safety and absence of risk to health in connection with the use, handling, storage and transport of articles and substances;
- provide such information, instruction, training, and supervision as is necessary to ensure the health and safety at work of employees;
- maintain any place of work under the employer's control in a condition that is safe and without risks to health and the provision and maintenance of means of access to and egress from it that are safe and without such risks;

- provide and maintain a working environment for employees that is safe, without risks to health and adequate as regards facilities and arrangements for their welfare at work;
- maintain records as required by health and safety legislation.

The duty of every employee

It is the duty of every employee:

- to take reasonable care for the health and safety of him or herself or others who may be affected by his or her acts or omissions at work;
- to cooperate with their employer in all matters relating to health and safety law and any duty or requirement that the employer may be required to make under that law.

It is also an offence for any employee to intentionally or recklessly interfere with or misuse anything provided in the interests of health, safety and welfare that the law may require.

12.3.3 Enforcement

The enforcement of the Health and Safety at Work Act, together with all other health and safety law, is carried out under the provisions of this main Act. This means that methods of enforcement are the same all over the country for all health and safety at work legislation.

The Health and Safety Executive are responsible for maintaining an inspectorate, and across the country there is a force of health and safety inspectors which consists of factory inspectors, agricultural inspectors, mines inspectors, nuclear inspectors and quarries inspectors.

The health and safety inspector's powers and duties

The health and safety inspector:

- may enter any premises at any reasonable time – or at any time (night or day) if they suspect a dangerous situation;
- can examine and investigate as necessary;
- can measure, take photographs, take samples or take possession of anything, if required for evidence;
- can inspect books and other documents that are relevant;
- can cause dangerous equipment to be rendered safe;
- can question people and require them to make a signed declaration of the truth of answers given;
- can call upon police to assist to enter premises if necessary;
- if there appears to be a contravention of safety law, may serve an Improvement Notice requiring certain work to be completed within a specified time. If it appears to be too dangerous to allow work to con-

tinue may serve a Prohibition Notice stopping the operation until remedial work is satisfactorily completed.

12.4 CONSTRUCTION (DESIGN AND MANAGEMENT) REGULATIONS 1994 (CDM)

It was recognized that the UK accident rate in construction was very high, resulting in more deaths than in any other industrial sector. As a consequence the Construction Industry Advisory Committee (CONIAC) worked together with the Health and Safety Executive (HSE) to develop proposals for new Regulations in the late 1980s.

The European Union reviewed the situation regarding construction safety across the member states and found it to be similar throughout. The results of this research was the Temporary and Mobile Construction Sites Directive adopted by the EC in June 1992. The Construction (Design and Management) Regulations were introduced by the HSE to implement this directive, the regulations coming into force on 31 March 1995.

The regulations supplement and update current UK legislation contained in the Health and Safety at Work Act 1974 and various regulations made under it. They apply to construction projects including demolition and everyone associated with them: owners, designers, contractors and subcontractors. The Regulations are about management of health and safety and while they do not apply to every construction project or everyone all of the time, most construction projects and the people working on such projects will be affected.

12.4.1 Scope and purpose

The main thrust of existing legislation prior to these regulations was to make individual employers responsible for the health and safety of their employees at work. These regulations, on the other hand, proceed on the basis that health and safety responsibilities in construction should be coordinated and be shared among all the parties involved, be they owner, designer, contractor or subcontractor.

The CDM regulations apply to construction work which is notifiable, i.e. lasts for more than 30 days or will involve more than 500 person days of work. CDM also applies to non-notifiable work which involves five people or more on site at any one time.

However, CDM applies to any design work no matter how long the work lasts and how many workers are involved on site. If the work includes demolition, CDM applies, regardless of the length of time or number of workers.

12.4.2 Duties

Five key parties (firms or individuals) have specific duties under the Regulations:

1. the owner;
2. the planning supervisor;
3. the designer;
4. the principal contractor;
5. contractors and subcontractors.

Duties of the owner

The owner's key duties are, as far as is reasonably practicable, to:

- select and appoint a competent planning supervisor, and principal contractor;
- be satisfied that the planning supervisor and principal contractor are competent and have the ability to allocate adequate resources for health and safety;
- be satisfied that designers and contractors are also competent, and will allocate adequate resources when making arrangements for them to work on the project;
- provide the planning supervisor with information relevant to health and safety on the project;
- ensure construction work does not start until the principal contractor has prepared a satisfactory health and safety plan;
- ensure the health and safety file is available for inspection after the project is completed.

Duties of the planning supervisor

The planning supervisor has to coordinate the health and safety aspects of project design and the initial planning to ensure that:

- designers comply with their duties, in particular, the avoidance and reduction of risk;
- designers cooperate with each other for the purposes of health and safety;
- a health and safety plan is prepared before arrangements are made for a principal contractor to be appointed;
- advice is given to the owner, as required, on the competence of proposed designers and contractors in health and safety matters, and as to whether the resources allocated by each of them to the project are adequate for health and safety purposes;
- the project is notified to the Health and Safety Executive;
- the health and safety file is prepared and delivered to the owner at the end of the project.

Duties of the designer

The designer's key duties are, as far as is reasonably practicable, to:

- alert owners to their duties;
- consider during the development of designs the hazards and risks which may arise to those constructing, operating and maintaining the structure;
- design to avoid risks to health and safety so far as is reasonably practicable;
- reduce risks at source if avoidance is not possible;
- consider measures which will protect all workers if neither avoidance nor reduction to a safe level is possible;
- ensure that the design includes adequate information on health and safety;
- pass this information on to the planning supervisor so that it can be included in the health and safety plan and ensure that it is given on drawings or in specifications, etc.;
- cooperate with the planning supervisor and, where necessary, other designers involved in the project.

Duties of the principal contractor

The principal contractor's key duties are to:

- develop and implement the health and safety plan;
- arrange for competent and adequately resourced contractors to carry out the work where it is subcontracted;
- ensure the coordination and cooperation of contractors and subcontractors;
- obtain from contractors the main findings of their risk assessments and details of how they intend to carry out high risk operations;
- ensure that contractors have information about risks on site;
- ensure that workers on site have been given adequate training;
- ensure that contractors and workers comply with any site rules which may have been set out in the health and safety plan;
- monitor health and safety performance;
- ensure that all workers are properly informed and consulted;
- display the notification of the project to the HSE;
- pass information to the planning supervisor for the health and safety file.

Duties of contractors and subcontractors

Contractors and subcontractors, including the self-employed acting as such, have duties to play their part in the successful management of health and safety during construction work. The key duties are to:

- provide information for the health and safety plan about risks to health and safety arising from their work and the steps they will take to control and manage the risks;

- manage their work so that they comply with rules in the health and safety plan and directions from the principal contractor;
- provide information for the health and safety file, and about injuries, dangerous occurrences and ill health;
- provide information to their employees.

The self-employed also have these duties when they act as subcontractors.

12.4.3 Health and safety documentation required under the Regulations

Under the Regulations a health and safety plan is required which should consist of the following.

Pre-tender health and safety plan

This document should be prepared in time so that it is available for contractors tendering to carry out or manage the construction work. It is the responsibility of the planning supervisor to ensure that the pre-tender plan is prepared and that it includes the following information:

- a general description of the work;
- details of timings within the project;
- details of risks to workers known or envisaged at that stage;
- information required by the potential principal contractor to demonstrate competence or adequacy of resources;
- information for preparing a health and safety plan for the construction phase and information for welfare provision.

Health and safety plan for the construction phase

This plan developed by the principal contractor is the foundation on which health and safety management of the construction work is based. It should include:

- arrangements for ensuring the health and safety of all who may be affected by the construction work;
- arrangements for the management of health and safety of construction work and monitoring of compliance with health and safety law;
- information about welfare arrangements.

Health and safety file

This is a record of information for the owner/end user, which tells those who might be responsible for the structure in future of the risks that have to be managed during maintenance, repair or renovation.

The planning supervisor has to ensure that it is prepared as the project progresses and it is given to the owner when the project is complete. The owner has to make it available to those who will work on any future design, building, maintenance, or demolition of the structure.

12.4.4 Enforcement

The primary responsibility for enforcing compliance lies with the Health and Safety Executive. It has powers to issue improvement and prohibition notices and in serious cases to bring criminal prosecutions. Currently breaches by the owner, planning supervisor, designer, principal contractor and subcontractors are punishable by a fine of up to £5000 in the magistrates' court or by an unlimited fine in the crown court.

Failure by the owner to ensure work does not start on site until the health and safety plan is prepared, or secondly failure by the principal contractor to keep unauthorized persons off the site, may result in a civil claim for damages.

12.5 REFERENCES

In order to be aware of current legislation it is advisable to consult the regular issue of the Monthly Advisory Service on Codes of Practice and the Changing Legal and Technical Requirements of Health and Safety at Work entitled *Health & Safety Monitor* published by Monitor Press. This reference publication provides up-to-date information on all health and safety issues, current legislation and reports of incidents and prosecutions.

The major Acts and Regulations pertinent to the construction industry are listed in Appendix B.

12.6 MAINTENANCE OF SAFETY RECORDS ON SITE

Statutory records which have to be kept on construction sites are numerous and varied and are required to meet a wide number of statutory instruments including many of those detailed in Appendix B. The records, which should be readily available for examination in site offices, can be split into two main types, i.e. those relating to plant and equipment and those relating to people.

The main record keeping requirements together with cross-references are detailed in Appendix C and although this list is not exhaustive it is indicative of the degree of record keeping necessary.

12.7 STATUTORY APPROVALS FOR CAPITAL PROJECTS

12.7.1 Scope and purpose

In the planning, design, construction and commissioning of any new plant a number of approvals must be obtained and certain notifications made under various Acts and statutory instruments. Many of these approvals and notifications are also necessary for modifications to or demolition of existing plants. These statutory requirements are administered by local authorities, statutory bodies and government departments.

The purpose of this chapter is to outline the proper procedures to be used by a project team in conjunction with the owner for obtaining the statutory approvals associated with new projects in England, Wales and Scotland. Changes in the legislation are frequently made and therefore the contents of this chapter should be used as a guide only, it being the responsibility of the project team to ensure that all the requirements of current legislation are met.

In most cases projects relate to sites where the owner is the landlord. Generally where one operating company is present on a site in the capacity of a tenant of another business or company and when projects encroach on land owned by third parties there will be additional consultation with the appropriate land owner.

A full understanding of statutory approval requirements is necessary to ensure:

- the business meets its legal obligations;
- an appropriate allowance is made in the financial budget for meeting statutory requirements;
- planning aspects of projects are completed quickly and cost effectively;
- good relations with the statutory authorities are developed and maintained;
- confidentiality of information disclosed to local authorities is preserved.

Exclusions

This chapter excludes the statutes and regulations that apply to:

- projects in foreign countries and Northern Ireland;
- statutory requirements which have to be met during plant operation (except where these must be taken into account during design);
- approvals in connection with specific mines and quarries legislation, railways and road vehicles.

12.7.2 Responsibilities

The owner project team must ensure that the project meets obligations with respect to the statutory requirements. Progress towards fulfilling these requirements should be reviewed regularly. The project team is responsible for ensuring that all statutory approvals are obtained for the project and that local deviations are complied with.

12.7.3 References

- Appendix D: example approvals and notifications to be considered for all projects form;
- Appendix D: example project record of actions form;
- Appendix E: the Local Government (Access to Information) Act 1985;
- Appendix F: environmental impact assessment (EIA).

12.7.4 Procedure

Preliminary action

In order to minimize costs of new projects, those undertaking negotiations with the statutory authorities should not propose or accept conditions more restrictive than are necessary to obtain the various approvals.Even if planning permission is not required the need for other statutory approvals shall be assessed.

Infringements of statutory provisions frequently arise over post-application additions or alterations to a new plant, or over modifications to an existing plant or building. It should not be assumed that the original statutory approval covers additions or modifications.

It is suggested that appropriate external advice on statutory requirements is sought for projects abroad, in connection with mines and quarries, or railways or road vehicles or relating to housing or agricultural property.

For international projects it is often necessary to obtain specialist expert advice.

Contacts with statutory authorities

During all contacts with local authorities, the provisions of the Local Government (Access to Information) Act 1985 should be borne in mind, as given in Appendix E.

At an appropriate stage in the life of a project the owner's project manager, or equivalent, should ensure that a meeting is convened to define the strategy for ensuring that all the approvals are obtained. Legal, safety and environmental departments should be consulted for advice on timing, etc., of initial contacts.

Discussions should take place with all the relevant statutory authorities in order to obtain their comments, advice and agreement in principle to the project. This ensures that when the planning application is submitted, delays are minimal when the planning authority sends out the application to these same statutory bodies for their approval.

The relevant statutory authorities should be informally kept aware of progress on the project up to the stage where their formal approval is needed.

Applications for planning permission should be approved and signed by the appropriate owner's nominee and should be supported by an environmental impact outline information brief as outlined in Appendix F. Certain types of projects also require an environmental impact assessment.

When considering major or potentially controversial planning applications, local authority planning departments commonly take the advice of external consultants. When approached by such consultants, care should be taken to ensure that:

- the role of the consultant is clearly understood by all parties;
- only such information as is relevant is released;
- the status of any information given to or discussed with the consultant in confidence is fully defined using Appendix E, The Local Government (Access to Information) Act 1985.

Other interested parties

Local action committees may employ consultants to plead their case with the local authority. Although it may be necessary to provide information for these non-official bodies in order to expedite the application, the greatest care should be exercised over its release and all of the project team should be aware of and involved in this exercise.

Monitoring progress

The project manager should maintain records of all the statutory consents, approvals and notifications that are applicable to the project and record discussions with the appropriate consultants and progress of the approvals being sought. An example of how these records can be maintained is shown in Appendix D.

The time scale involved in obtaining all the necessary statutory approvals will vary considerably from project to project. If a project involves operations outside the site boundary, such as cross-country pipelines or diversion of rights of way, years rather than months may have to be allowed.

12.8 CONCLUSION

As can be seen from the foregoing, health, safety and environmental issues are complex and a lack of understanding can lead to serious consequences for both the project and more importantly the people involved in the project. It is therefore imperative that the project team identify all the requirements of current legislation, make allowances both in the project programme and financial budget for them and ensure that they are fully implemented.

Appendix A
Glossary of terms

A book of this size and complexity requires a chapter which enables the terms used to be detailed and defined for easy recognition. This glossary of terms is presented in alphabetical form for ease of use. It attempts to cover the majority of the terms referred to in the document but reference is made to the following two publications which cover a wider sphere of terminology and which have been used as a guide in the preparation of this glossary of terms:

- The Association of Cost Engineers *Cost Engineering Terminology*
- British Standards Institute's Documentation on Projects, Planning and Contracts

Account	Primary grouping of costs established for administrative and accounting purposes
Accountability	Defined level or point of budget authority and control responsibility
Account code	Code used to identify an individual cost account, the code may be alpha or numeric or a combination of the two
Accounts payable	Value of goods and services received for which payment has not yet been made
Accounts receivable	Value of goods and services executed for which payment has not yet been received
Accrual	Costs incurred but not yet invoiced and recorded in the accounts
Accrual basis	System of accounting records based on cost incurred and income earned
Accrued cost	Total recorded cost plus accruals
Accrued income	Total recorded income plus accruals
Accuracy factor	Factor used in estimating to indicate the likely range of accuracy of an estimate
Activity	An operation or process consuming time
Activity-on-arrow network diagram	A network in which the arrows symbolize the network activities
Activity-on-node network precedence diagram	Network in which the nodes symbolize the network activities
Advanced funding	Provision of funds, to cover future payments for goods or services purchased for and on behalf of an owner
AFC	(1) Approved for construction; (2) anticipated final cost
AFD	Approved for design

AFE	Authorization for expenditure
Ageing rate	Predetermined assessment of depreciation expense or capital cost relative to a period of time
Ancillaries	Necessary auxiliary components, usually unspecified, particularly of plant and equipment
Appropriation	Authorized allocation of funds for specified future work
Appropriation estimate	Early cost estimate, usually by owner, prepared from preliminary but largely fixed design data to an accuracy of -15% to +25%, for early budget purposes to establish viability of proposed project and to secure necessary funding approvals, also known as class III, evaluation, outline definition, sanction, study or scope estimate
Appropriation request	Documented application for future project funding requirements, prepared for management approval
Arrow	A directed connecting line between two nodes in a network
Assets	Current value of property, investments, stock and cash
Backcharge	Claim made against supplier, contractor or subcontractor for remedial work, loss or damage effected in respect of goods or services
Ballpark estimate	*See* Order of magnitude estimate
Bar chart	A chart on which activities and their durations are represented by lines drawn to a time scale
Bar chart schedule	Graphical representation of project activities or resources drawn as horizontal lines against a common time scale
Base cost	A cost established as a basis for calculating escalation, changes, provisions or other related variance
Base date	A date established as a basis for calculating price variations, forward escalation, currency differentials or other time-based cost data
Base estimate	An estimate prepared against a defined scope of work and established as a basis for evaluating subsequent adjustments, modifications or changes
Base exchange rate	A currency exchange rate used in preparation of a cost estimate or budget and established as a basis for calculating subsequent actual currency differentials or variances
Basic price	Standard rate or value adjustable by discount, escalation or variation to suit specific circumstances
Battery limits	Line of demarcation defining the extent of project scope, on-site facilities or contractual responsibility
Bid analysis	Activity of analysing suppliers' quotations or contractors' tenders
Bid bond	Sum provided by bidder as guarantee that bid will be submitted, or that if accepted, bidder will proceed into the contract

Bill of materials	List of material quantities estimated or measured from drawings for estimating or procurement purposes
Bill of quantities	List of construction quantities estimated or measured from drawings for tender pricing and contract payment purposes
Bill of quantities contract	Contract awarded against a bill of quantities for which payment is contract made in accordance with the unit prices contained in the bill of quantities
Bond	Sum of money, securities or a guarantee of a third party to guarantee completion of the work and/or recover sums payable or owed by the contractor under the terms of contract
Bonus	Incentive payment over and above agreed wage or contractual reimbursement, usually relating to achievement targets
Bonus incentive/bonus scheme	System agreed to increase output from work scheme force by promised payments in respect of identified and tabulated targets for achievement
Bonus/penalty clause	Contractual incentive placed on supplier or contractor to achieve certain completion targets for which a bonus payment is earned or a financial penalty incurred
Bonus/penalty contract	Contract containing a bonus/penalty clause
Book value	Asset value as recorded in the books of a business
Budget	(1) Authorized target for project achievement, or part thereof, usually expressed in terms of quantities, man-hours or costs; (2) a financial and/or quantitative statement, prepared and approved prior to a defined period of time, of the policy to be pursued during that period for the purpose of attaining a given objective
Budget change	Authorized change to previously established budget
Budget cost	Authorized target costs for completion of the work
Budget estimate	Semi-detailed estimate prepared for budget purposes from preliminary but largely fixed design data to an accuracy of -10% to +15%. Also known as class II, preliminary, sanction, scope or semidetailed estimate
Budgetary control	The establishment of, and the continuous comparison of actual with budgeted results, either to secure the objective of that policy individual or to provide a basis for its revision
Building services	Mechanical, electrical or other specialized service facilities existing or to be installed within a building
Bulk factors	Multipliers used to raise manufactured equipment costs to erected plant or total installation costs, inclusive or associated bulk materials and ancillaries
Bulk materials	General term for any material commonly purchased in bulk
Burden	Overhead expenses distributed as an add-on cost to direct cost

Cancellation charges	Charges submitted by supplier or contractor in respect of cancellation of purchase order or termination of contract by owner or others
Capacity factor	Term used in estimating in respect of the capacity ratio between two similar plants
Capital	Money invested or available for investment against which earnings and dividends are assessed
Capital cost estimate	Any cost estimate for a future fixed asset, including design, supply and installation costs, but excluding any finance or operating costs
Capital expenditure	Expenditure attributable to the creation of a permanent or fixed asset
Capital recovery	Recovery of capital invested in a project over the life of that project, including allowed depreciation
Cascade chart	A bar chart on which the vertical order of activities is such that each activity is dependent only on activities higher on the list
Cash	Moneys in hand or available for immediate use
Cash basis	Method of recording cost and income data based on actual cash payments and receipts
Cash call	Projection and request for future cash requirements at specific points in time
Cash flow	Projected cash balance over a stated time period based on the net flow of actual or anticipated cash payments
Change estimate	Estimate prepared for evaluation of a potential change to the project
Change impact	Effect of change on project cost and schedule
Change order	Document authorizing execution of a defined project change
Change of scope	Any deviation from a previously established scope of supply or scope of work
CIF	Carriage, insurance and freight
Claim	Demand made for compensation or additional remuneration to which one of the parties to the contract considers he has a contractual right
Class I estimate	*See* Definitive estimate
Class II estimate	*See* Budget estimate
Class III estimate	*See* Appropriation estimate
Class IV estimate	*See* Order of magnitude estimate
Close-out report	Final end-of-project or end-of-assignment report
Code of accounts	Complete set of account codes developed for the purpose of cost identification, cost allocation and cost reporting, as necessary for cost control or future estimating purposes
Commitment	Total value of all orders placed to date, including letters and telexes of intent, and including allowances for all additional costs anticipated in respect of such orders
Committed cost	Costs or liability that are debited to the project budget

Conditions of contract	Terms agreed between the parties for the execution of a contract
Construction indirects	Term commonly used to describe all indirect field expenses, such as non-productive labour, field staff, field office staff, construction tools, etc.
Construction phase	Period within the life of a project covering the site construction activity
Construction philosophy	Overall plan of construction work, stating intended methods of execution, sequence and timings of key activities, utilization of direct or subcontract labour resources, etc.
Consumables	Material consumed or construction aids permanently incorporated into a construction project, including non-recoverable formwork, shims and packings, welding electrodes, industrial gases, fuel, etc.
Contingency	Budgetary provision for unforeseeable occurrences within a defined project scope
Contingent reserve	Current value of contingency held or remaining
Contract	Legal and binding agreement in writing between two or more parties, defining scope of work or services to be provided, terms and conditions and financial settlement
Contract period	Time stated in contract documents for the execution of the works
Contract price adjustment (CPA)	Adjustment to the contract sum, usually in respect of escalation and for which provision is made in the contract
Control budget	Any budget used for control purposes, against which actual performance is measured
Control estimate	Any estimate used as a basis or yardstick for cost control purposes
Cost	Expense, estimated or actual
Cost analysis	Systematic breakdown of cost data into elements or categories for detailed examination
Cost breakdown structure (CBS)	Systematic breakdown of costs into predefined elements in order to summate cost of like items
Cost centre	Principal cost collection facility established for budgetary control
Cost code	Identifying code assigned to specific cost items in accordance with a defined, methodical and usually hierarchical classification system
Cost control	Systematic restraint on expenditure in order to secure completion within predetermined budgets or targets
Cost engineer	An engineer who by virtue of his training and experience is competent to develop and make practical use of the principles of engineering cost management
Cost engineering	That area of engineering practice devoted to the problems of project cost management, involving such activities as estimating, cost control, investment appraisal and risk analysis

Cost estimate	Any considered prediction of future or final project costs
Cost index	Factored tabulation of recorded cost movements for a particular category or item or unit of work over a specified period of time as measured from a defined base date
Cost indices	Series of factored tabulations as defined above for cost index
Cost monitoring	Close observation of the costs of work in progress with reference to the estimated cost targets
Cost norm	Established average or expected cost of defined item or unit of work
Cost overrun	Any cost in excess of that budgeted
Cost plus contract	General term describing a form of contract in which reimbursement is based on actual costs incurred plus a mark-up or fee to cover overheads and profit
Cost report	Formal record or register of current cost information prepared in an agreed manner for reporting, statistical or record purposes
Cost to date	Any costs variously committed, incurred, expended, accrued, etc. to a specifically defined current cut-off date
Cost underrun	Any cost less than budgeted
Cost variance	Confirmed or forecast deviation between actual or forecast cost and previously established budget or estimate
CPA	(1) Contract price adjustment; (2) critical path analysis
Critical activity	An activity on a critical path
Critical event	An event on a critical path
Critical path	A path through a network with least float
CTR	Cost, time and resource – particular concept of project control based on assigning costs, durations and resources to individual network activities
Currency differential	Cost variance due to currency exchange rate fluctuations
Daywork	Work paid for on an hourly, daily or weekly time basis covering labour, material and plant charges
Definitive estimate	Detailed estimate prepared for control purposes from well defined design data to an accuracy of −5% to +10%. Also known as class I, execution, detailed or final estimate
Deliverables	Term used to describe documents, materials and other physical commodities capable of being 'delivered' on a project in accordance with a specific programme or schedule
Demurrage	Charges payable for delay or detention of delivery
Depreciation	Fall in value of a capital asset due to use and/or lapse of time

Design and construct contract	Contract based on a brief provided by the owner in which a contractor designs and constructs a project
Design basis	Formal front-end document defining project scope of work and anticipated execution philosophy as required for engineering design activity, and upon which estimated costs of the project are based
Design cost	Cost of preparing defined design, inclusive of all fees, drawings, specifications, models, surveys, and all other documentation or services necessary to a completed package
Design development allowance	Nominal allowance included in an estimate or forecast to cover additional costs resulting from natural development of preliminary design data
Detailed estimate	*See* Estimate (class I)
Direct man-hours	(1) Man-hours which can be directly allocated to a productive account or cost centre, as opposed to indirect man-hours; (2) man-hours in respect of labour directly employed by owner or principal, as opposed to man-hours for contract labour
Duration	The estimated or actual time required for the completion of an activity
Errors and omissions excepted (E & O E)	A qualification clause commonly used to cover mistakes in quotations
Earliest event time (EET)	The earliest time by which an event can occur within the logical and imposed constraints of the network
Earned man-hours	Budget man-hours for work completed
Earned value	Budget value for work completed
EP & C	Engineering, procurement and construction – term commonly used in respect of contract services provided
Estimate	Method of categorizing estimates within particular levels of detail classification and expected accuracy (e.g. class 1, class 11 and class III estimates)
Evaluation estimate	*See* Appropriation estimate
Event	A stage in the progress of a project after the completion of all preceding activities but before the start of any succeeding activity
Event time	The time by which an event can be (or is to be) achieved
Expenditure	Value of all invoices and charges paid or approved for payment
Factored estimate	(1) Preliminary estimate prepared by factoring known costs from a previous similar project; (2) preliminary or budget estimate prepared by factoring the costs of major items in order to establish the cost of minor or bulk components
Final account	Final agreed settlement of payment under a contract

Final measure	Final measurement from as built drawings or site construction
Firm price	Price fixed in all respects except perhaps escalation, for which additional payment may be made in accordance with a stated formula and base date
Fixed lump sum price	Single price for specified scope of work not subject to adjustment for escalation or other reasons
Fixed price	Price for specified goods or services which remains unalterable for specified duration
Float	The time available for an activity or path in addition to its duration (it may be negative)
Fluctuation clause	Condition inserted into the terms of a tender or contract stating the method of dealing with price adjustments due to cost of living increases during the contract period
Free on board (FOB)	Cost of goods plus all charges until placed on board a vessel at a named port
Force majeure clause	Clause freeing the contractor from specific obligations in the event and as a result of an Act of God, such as earthquake, flood, war, riot, etc. as may be specifically detailed in the conditions of contract
Free issue material	Fabrication or construction material supplied or made available by owner or design contractor at no cost to fabricator or construction contractor
Hammock	An activity, joining two specified events, that may be regarded as spanning two or more activities. Its duration is initially unspecified as it is only determined by the difference between the earliest and latest times of the events concerned
Histogram	Vertical bar chart indicating estimated or actual manpower over each period of the contract
Incentive payments	Additional payments to labour for incentive purposes
Incurred costs	Actual cost to date of services rendered, work performed or materials supplied
Indirect costs	Costs not directly allocated to a productive account
Indirect man-hours	Man-hours which cannot be directly allocated to a productive account
Indirects	General term commonly used to describe all labour, material, plant, tools, supplies, services, etc. which cannot be directly allocated to a productive account
Indefinite schedule	Unlimited schedule — a schedule produced without resource constraint
Interface	An activity or event common to two or more events
Job card	Printed card or computer generated document authorizing specific elements of site work to be carried out in accordance with stated schedule and budget data

Key event	An event or milestone selected for its importance in the project
Labour only contract	Contract for supply of labour only
Ladder	A device for representing a set of overlapping activities where the start and finish of each succeeding activity are linked only to the start and finish of the preceding activity by lead and lag activities which consume only time
Lag	The minimum necessary lapse of time between the finish of one activity and the finish of an overlapping activity
Lang factors	Series of factors used in estimates to convert the cost of manufactured plant into erected or installed plant cost and, ultimately, completed project cost
Latest event time (LET)	The latest time by which an event has to occur within the logical and imposed constraints of the network, without affecting the total project duration
Lead	The minimum necessary lapse of time between the start of one activity and the start of an overlapping activity
Liquidated damages	Predetermined daily or weekly amount recoverable from a contractor in the event of delay in completion
Loop	An error in a network which results in a later activity imposing a logical restraint on an earlier activity
Lost time	Work time lost through labour problems, equipment failure, inclement weather or other cause
Lump sum	Single total price for work, goods or services covered by agreed specification or drawings
Managing contractor	Prime contractor responsible to owner for overall management of project
Man-hour forecast	Forecast of man-hours to be expended
Man-hour norm	Standard duration for undertaking a task
Master network	A network showing the complete project, from which more detailed networks are derived
Material take-off (MTO)	(1) Process of measuring material quantities from drawings; (2) document resulting from the process of measuring material quantities from drawings
Measured term	Term contract based on a schedule of rates, with payment based on valuing work done at the rates quoted
Measured work	Construction or fabrication work which is identified and measured, either physically or from drawings, for tender, progress or payment purposes
Mechanical completion	Recognized milestone for the completion of a process plant project, from which point commissioning can commence
Method statement	Statement in contract indicating general method of working

Milestone	Principal or critical event or occurrence within a project programme
Mobilization	Term used to describe preparatory setting-up activity
Monte Carlo simulation	Particular method of computerized probabilistic determination commonly utilized in risk analysis procedures
Network (project network)	A representation of activities and/or events with their interrelationships and dependencies
Network analysis	System of resource measurement and control relative to a planned schedule
Node	A point in a network at which arrows start and/or finish
Non-productive time	Paid labour time with no productive output, such as meal time, walk time, lost time, overtime uplift, etc.
Non-splittable activity	An activity that, once started, has to be completed without interruption or removal of resources for use on other activities
Norm	Standard cost or duration for specific task
OCPCA norms	Standard man-hour norms for mechanical and electrical work developed and published by Oil and Chemical Plant Contractors Association
Offsites	Term relating to areas of work outside battery limits of the main process plant and usually comprising delivery, product storage, administration and utilities such as water treatment, sewage treatment, etc.
Order of magnitude estimate	Term commonly used for very preliminary cost estimate Also known as class IV, ballpark, guesstimate, horseback, seat of the pants, screening estimate
Overhead costs	Company expenses not directly allocatable or chargeable to a particular project
Overtime	Hours worked in excess of defined normal times
Owner	In this document the word owner is used as a common term covering employer, purchaser, manufacturer, client or operator
Path	An activity or unbroken sequence of activities
Payroll burden	Indirect payroll costs, such as national insurance, holidays, etc.
Performance bond	*See* Surety bond
PERT	Programme evaluation and review technique
P & I D	Piping and instrument diagram
Precedence matrix	A table of activity identification used to show the sequence and interrelationship of activities in a network
Premium time	Overtime paid at uplifted rate
Price variation formula	Formula for evaluating contract prices to reflect change in economic conditions

Prime cost	Basic cost of installed plant, materials, labour or services, exclusive of indirect charges, operating costs, maintenance costs or loan interest
Prime cost sum (PC sum)	Sum included in a contract for an item or work to be supplied or carried out by a supplier of subcontractor and excluding the attendance and profit margin of the general contractor
Probability	Statistical concept which takes account of possible variations among a large number of occurrences or estimates
Probability analysis	Term given to systematic evaluation of probability as used in risk analysis
Productivity	Measure of efficiency of productive work relative to an established base or norm
Progress payment	Payment made to supplier or contractor in respect of progress
Progress weighting	Method of relating different activities in respect of a common base, in order to achieve an overall progress measurement
Project control	The proactive setting and monitoring of targets, analysis of performance, identification and anticipation of inefficiencies and implementation of preventive or remedial actions
Project life	Overall project duration, from inception to completion
Project manager	The individual to whom authority, responsibility and accountability has been assigned for the overall management of resources (including the technical, time and cost aspects) of a project, and the motivation of all those involved (see 2.3 of BS 6046: Part 1: 1984)
Project management	The mobilization and management of resources (including the technical, time and cost aspect) for the purpose of completing a project (see 2.3 of BS 6046: Part 1: 1984)
Project management team	A multi-disciplinary group of people headed by a project manager established to fulfil the range of managerial responsibilities inherent in the project management task (see 2.3 of BS 6046: Part 1: 1984)
Project monitoring	The comparison of current project status with that planned to identify and explain any deviations
Project programme	A diagram or list showing work to be done, with associated time scales
Provision	Allowance in estimate or forecast for a specific item or occurrence which cannot be fully defined
Provisional sum	Specified sum allowed in contract for item of work or supply which cannot be fully defined
Radius allowance	Allowance paid to site construction staff to cover local transport, etc.

Reimbursable contract	General term used to describe any contract in which payment is made on a cost reimbursement basis, either against rates, at cost plus a mark-up, or at straight cost
Resource cumulation	The process of accumulating the requirements for each resource to give the total required to date at all times throughout the project
Resource levelling	The process of producing a schedule that reduces the variation between maximum and minimum resource requirements
Resource limited scheduling	The scheduling of activities, so that predetermined resource levels are never exceeded
Resource smoothing	The scheduling of activities, within the limits of their float, so that fluctuations in individual resource requirements are minimized
Retention	Sum of money retained under contract for specified period of time in respect of possible defective work or service
Risk analysis	Process of systematic evaluation of risk possibilities in respect of uncertain project outcome
Royalty	Payment for use of copyrights and patents, such as for proprietary chemical processes
Schedule of rates	List of unit prices for work to be undertaken and/or materials to be supplied
Schedule of rates contract	Open-ended contract with reimbursement against a schedule of rates
Scope of supply	Defined extent of services, commodities or facilities to be provided under contract
Scope of work	Defined extent of work to be executed under contract
'S'-curve	Graphical control tool commonly used for monitoring cost, progress or resource performance against time
Skeleton network	A summary network obtained by reducing the number of activities in a network without changing the logical or timing relationships of the remaining activities
Slack	The calculated float time within which an event has to occur within the logical and imposed constraints of the network, without affecting the total project duration. Note 1: It may be made negative by an imposed date. Note 2: The term slack is used as referring only to an event
Splittable activity	An activity that can be interrupted in order to allow its resources to be transferred temporarily to another activity
Star rate	Agreed additional unit rate for work not covered by the original contractual schedule of rates
Summary network	A network in which the amount of detail presented is condensed

Supplier	Company supplying material or specialist equipment to a project. Supplier's work may include supervision of installation and commissioning
Surety bond	Legal document executed by the owner with a third party to guarantee that a contractor performs its contracted liability
Tender	Written formal offer stating price, completion time and any qualifications submitted for the purpose of entering into a contract
Term contract	Contract that enables a client to order work at pre-agreed rates during a prescribed period
Tied activities	Activities that have to be performed sequentially or within a predetermined time of each other
Time-based network linked bar chart	A bar chart that shows the logical relationships between activities
Time limited scheduling	The scheduling of activities, so that the specified project duration, or any imposed dates, are not exceeded
Total float	The time by which an activity may be delayed or extended without affecting the total project duration
Turnkey contract	Contract for design, supply, construction and commissioning of a plant or facility
Unit cost	Cost per stated item or unit of measurement
Unlimited schedule	A schedule produced without resource constraint
Valuation	Assessment or calculation of completed work for payment purposes
Value	The worth of activities in terms of budgeted man-hours or costs
Value engineering	Technique of ensuring completion at most economic cost, without loss of quality
Variation	Change in scope of work or supply, subject to confirmation and approval
Variation order	*See* Change order
Work breakdown structure	The way in which a project may be divided into discrete groups for programming, cost planning and control purposes (*see* also Work package)
Work package	A group of activities (normally related in some way or ways) that are defined at a level within a work breakdown structure
Yardstick	Historical data established for measurement, evaluation and control of current work

Appendix B
Acts and Regulations pertinent to the construction industry

PRINCIPAL LEGISLATION

The Public Health Act 1936 (specifically that part of the Act covering the construction of premises used for the purposes of a food business and their repair and maintenance)
The Factories Act 1961
Town and Country Planning (Assessment of Environmental Effects) Regulations 1988
Town and Country Planning (Environmental Assessment – Scotland) Regulations 1988
The Fire Precautions Act 1971
The Health & Safety at Work Act 1974
The Control of Pollution Act 1974
The Construction (Design and Management) Regulations 1994
The Food and Environment Protection Act 1985

SUBORDINATE LEGISLATION

Petroleum (Consolidation) Act 1928
Occupiers' Liability Act 1957
The Work in Compressed Air Special Regulations 1958
Radioactive Substances Act 1960
The Construction (Lifting Operations) Regulations 1961
The Construction (General Provisions) Regulations 1961
Offices, Shops and Railway Premises Act 1963
Construction (Notice of Operations and Works) Order 1965
The Construction (Health & Welfare) Regulations 1966
The Construction (Working Places) Regulations 1966
Employers' Liability (Defective Equipment) Act 1969
Employers' Liability (Compulsory Insurance) Act 1969
Abrasive Wheels Regulations 1970

The Highly Flammable Liquids & Liquefied Petroleum Gases Regulations 1972

Gas Safety Regulations 1972

Employment of Children Act 1973

The Protection of the Eyes Regulations 1974

The Woodworking Machines Regulations 1974

The Health & Safety at Work Act 1974 (Application outside Great Britain) Order 1989 (specifically covering work in connection with offshore installations and pipelines)

The Safety Representatives and Safety Committee Regulations 1977

The Control of Pollution (Special Waste) Regulations 1980

The Safety Signs Regulations 1980

Control of Lead at Work Regulations 1980

Industrial Diseases (Notification) Act 1981

The Diving Operations at Work Regulations 1981

The Health and Safety (First Aid) Regulations 1981

Notification of Installations Handling Hazardous Substances Regulations 1982

Notification of New Substances Regulations 1982

The Asbestos (Licensing) Regulations 1983

Control of Industrial Major Accident Hazards Regulations 1984

The Reporting of Injuries, Diseases and Dangerous Occurrences Regulations 1985

The Ionizing Radiations Regulations 1985 (to be observed where NDT is required)

Control of Asbestos at Work Regulations 1987

Consumer Protection Act 1987

Petroleum Act 1987

Collection and Disposal of Waste Regulations 1988

Dangerous Substances (Notification and Marking of Sites) Regulations 1988

The Control of Substances Hazardous to Health Regulations 1988 (COSHH)

The Control of Pesticides Regulations 1986

The Asbestos at Work Regulations 1989

The Electricity at Work Regulations 1989

The Noise at Work Regulations 1989

Pressure Systems and Transportable Gas Containers Regulations 1989

Food Safety Act 1990

Environmental Protection Act 1990

Planning (Hazardous Substances) Act 1990

Lifting Plant and Equipment (Records of Test and Examination, etc.) Regulations 1992

Management of Health and Safety at Work Regulations 1992

Manual Handling Operations Regulations 1992

Notification of Cooling Towers and Evaporative Condensers Regulations 1992

Personal Protective Equipment at Work Regulations 1992

Provision and Use of Work Equipment Regulations 1992

Workplace (Health, Safety and Welfare) Regulations 1992

The Plant Protection Products Regulations 1995

Appendix C
Safety records required on site

PLANT AND EQUIPMENT ISSUES

	Form
Scaffolding weekly inspections	F91 (Part 1)
General access	F91 (Part 1)
Tower scaffolds	F91 (Part 1)
Suspended access equipment	F91 (Part 1)
Lifting appliances	
Cranes	F96
	F91 (Parts 1 & 2)
	F75
Hoists	F91 (Part 1)
	F92 (Part 2)
Excavators (used as crane)	F91 (Part 1)
Other lifting appliances	F80
	F91 (Part 1)
	F9 (Part 2)
Lifting gear	F87
	F97
	F91 (Part 1)

Portable electrical equipment
  HSE Guidance Note PM32 (1983) Electricity at Work Regulations require regular inspection and documentation of results.

Pressurized systems (air compressors, etc.)
  Records of periodic inspections need to be maintained as required by Pressurized Systems Regulations.

Excavations
  Excavations need to be regularly examined (Form F91 Part 1 Section B).

ASSESSMENTS

The following assessments need to have been carried out:
Control of substances hazardous to health
Manual handling
VDU operators
Risk assessment
Form C9 to be issued for identification of hazardous work

PEOPLE ISSUES; LISTS OF NOMINEES OR APPOINTEES REQUIRED

Abrasive wheels
Asbestos workers
Competent persons for inspecting scaffolds
Competent persons for inspecting lifting tackle
Competent persons for inspecting electrical gear
First aiders
Dumper/excavator/crane drivers, etc.
Competent persons for ionizing radiations
Permit to work system where required should be in place and records held.
Competent persons for issuing
Competent persons for issuing entry permits as required under Chemical Works Reg. 7 (1922)
Accident book B1510 and F2508/F2508A as required for RIDDOR incidents/accidents

Appendix D
Example forms

APPROVALS AND NOTIFICATIONS TO BE CONSIDERED FOR ALL PROJECTS

PROJECT NO: Date: Rev: Rev Date:	BUSINESS AREA:	PROJECT TITLE:	
APPLICATION FOR/ OR NOTIFICATION OF	STATUTORY AUTHORITY TO WHOM APPLICATION/ NOTIFICATION SHALL BE MADE	RESPONSIBLE FOR NEGOTIATION WITH STATUTORY AUTHORITY	BUSINESS GROUP/ COMPANY FUNCTIONS TO BE INVOLVED
Relating to sites, plants and buildings			
Planning permission	Local planning authority		
Environmental Impact	Local planning authority		
Building Regulations Approval	Local authority		
Pharmaceutical production	DoH Medicines Inspectorate		
High buildings and structures	Local planning authority Civil Aviation Authority		
Registration of Factories and Offices	Health and Safety Executive (HSE)		
Diversion of rights of way	Local authority or magistrates court		
Marine and waterway development	Port Authority		
Works in or by rivers or canals	National River Authority British Waterways Board		
Fire certificate	HSE Local authority/local authority building inspector		
Relating to handling and storage of materials			
Installations handling hazardous substances	Local planning authority		
Control of Industrial Major Accident Hazards (CIMAH)	HSE		
Hazardous Substances Consent Regulations	Local authority		
Licence to store petroleum spirit	HSE Local fire authority		
Approval to store highly flammable liquids and LPG	HSE		
The Installation of a refinery or bonded warehouse	HM Customs and Excise		
Certificate for Storage of Explosive Material	Police		
Authorization for cross country/local pipelines	Dept of Energy Pipeline Inspectorate		

Certificate of weighing or measuring
  equipment
	Trading Standards
	Weights & measures inspector
	  (HSE)		

New substances	Employment Medical Advisory Service		
Storage of special waste	Waste Regulation Authority		
Control of Substances Hazardous to Health (COSHH)	HSE		
Registration of Radioactive Sources	Her Majesty's Inspectorate of Pollution (HMIP). Radiochemical Inspectorate		

APPLICATION FOR/ OR NOTIFICATION OF	STATUTORY AUTHORITY TO WHOM APPLICATION/ NOTIFICATION SHALL BE MADE	RESPONSIBLE FOR NEGOTIATION WITH STATUTORY AUTHORITY	BUSINESS GROUP/ COMPANY FUNCTIONS TO BE INVOLVED
Relating to environmental control			
Authorization to operate a prescribed process (EPA)	HMIP or local authority		
Licence for water abstraction	National Rivers Authority		
Certificate for a reservoir	Local authority (county)		
Consent to discharge to underground cavities	Waste Regulation Authority National Rivers Authority HSE		
Consent to discharge liquid effluents into controlled waters	National Rivers Authority		
Consent to discharge liquid effluents into public sewers	Sewerage undertaker/local authority		
Licence for disposal of waste on land	Waste Regulation Authority		
Emissions to atmosphere	HMIP Local authority (Environmental Health Dept)		
Consent under noise control regulations	Local authority		
Relating to health and safety of people during construction			
Start of construction or demolition	HSE		
Licence for disposal of construction waste	County waste regulation authority		
Operations involving asbestos	HSE		
Registration of radioactive sources	HMIP Radiochemical Inspectorate		
Fire certificate for construction offices	HSE Local authority		
Notification of use of construction offices	HSE		
Transport of abnormal loads	Local authority Police		
Work near overhead power lines	Electricity company		
Relating to notification before start-up			
Fire certificate	HSE Local fire authority		
Notification of cooling tower	Local authority		
Building Regulations (periodic site visits) Notification to HSE	Local authority HSE		

PROJECT RECORD OF ACTIONS FORM

PROJECT NO: Date: Rev: Rev Date:	BUSINESS AREA:		PROJECT TITLE:		
STATUTORY REQUIREMENT	IS THIS A STATUTORY REQUIREMENT FOR THIS PROJECT?	DATE ON WHICH AGREEMENT IN PRINCIPLE REACHED	FORMAL APPLICATION/ NOTIFICATION SUBMITTED	FORMAL APPROVAL RECEIVED	DEPT WHERE LICENCE/ CONSENT IS FILED
Relating to sites, plants and buildings					
Planning permission			(3)		
Environmental impact statement outline					
Building Regulation approval Each geographically separated unit (a) Amenity (b) Process (c) Storage			(3)		
Approval for pharmaceutical production					
Agreement on high buildings and structures		(2)			
Registration of Factories and Offices					
Diversion of rights of way					
Agreement for marine/waterway development					
Agreement for work in or by rivers or canals					
Fire certificate (a) Amenity/office space (b) Special premises		(2)			
Relating to handling and storage of materials					
Notification of installations handling hazardous substances					
Control of Industrial Major Accident Hazards (CIMAH Regulations) (a) Preparation of safety case (b) Preparation of onsite emergency plan (c) Provision of information for offsite emergency plan (d) Unless already done by site, information to persons liable to be affected		(2)			
Hazardous substances consents					
Licence to store petroleum spirit			(3)		
Approval to store and handle highly flammable liquids					
Agreement on the installation of a refinery or bonded warehouse					
Certificate for storage of explosive materials					
Authorization for pipelines (a) Cross-country (b) Local		(2)			
Certification of weighing instruments					
Notification of new substances					
Licence for storage for special waste					
COSHH assessment					
Registration of radioactive sources					

STATUTORY REQUIREMENT	IS THIS A STATUTORY REQUIREMENT FOR THIS PROJECT?	DATE ON WHICH AGREEMENT IN PRINCIPLE REACHED	FORMAL APPLICATION/ NOTIFICATION SUBMITTED	FORMAL APPROVAL RECEIVED	DEPT WHERE LICENCE/ CONSENT IS FILED
Relating to environmental control					
EPA Part A/Part B					
Licence to abstract water					
Certificate for a reservoir					
Consent to discharge to underground strata		(2)			
Consent to discharge to controlled waters		(2)			
Consent to discharge to public sewers		(2)			
Licence for disposal at sea		(2)	(3)		
Licence for disposal on land					
Approval of emissions to atmosphere					
Consent under noise regulations					
Relating to health and safety of people during construction					
Start of construction or demolition					
Licence for disposal of construction waste					
Operations involving asbestos					
Registration of radioactive sources					
Fire certificate for construction offices					
Notification of use of construction offices					
Transport of abnormal loads					
Work near overhead power lines					
Relating to notification before start-up					
Fire certificate examinations					
Notification of cooling tower					
Building regulations (periodic site visits)					
Notification to HSE					

Notes:

(1) During HSE Study, indicate yes or no, or raise an action for the project manager to consult specialists for an authoritative answer.

(2) If this approval is necessary and if planning permission is required then agreement in principle should be reached before submission of the planning application.

(3) Projects which do or may require an EIA should be discussed with the local authority planning department to determine the need for, and extent of, any assessment of environmental impact.

Revision
Date
Approved by (project manager)

Appendix E
The Local Government (Access to Information) Act 1985

SCOPE AND PURPOSE

This Act extends to England, Wales and Scotland and provides for greater public access to local authority meetings, reports and background documents. Its provisions shall be taken into account in relation to any document containing sensitive information which is to be passed to local authority.

Commercially sensitive information disclosed orally to a local authority officer or councillor is not affected by the Act unless made the subject of:

- a report by an Officer which is submitted to a meeting;
- a background paper to a report

The Act applies to council, committee and sub-committee meetings of all local authorities, but does not apply to parish and community councils.

INFORMATION GUIDANCE

The Act requires that meetings shall be open to the public except:

- where 'confidential information' is likely to be disclosed (in which case there is an obligation to exclude the public), or
- where 'exempt information' is likely to be disclosed and the meeting for that reason decides to exclude the public. (Where exempt information is concerned, the meeting has a discretion whether or not to exclude the public.)

The means to safeguard 'exempt information' are likely to be unreliable and *it should be assumed therefore that any other document passed to any local authority may become public*. Indicate on any confidential document or part of it which it is proposed to pass to any local authority, the following:

- this document (or section/para/page xyz) contains 'exempt information' for the purposes of the Local Government (Access to Information) Act 1985 and shall not be disclosed to the public without authorization from the owner (name originator);

- information discussed with or passed to a technical consultant acting on behalf of a local authority should be the subject of a secrecy agreement. The project manager shall ensure that consultant's reports do not contain 'exempt information'.
- the business should be consulted in connection with any secrecy agreement and the passing of confidential and/or 'exempt information' to any local authority.

GLOSSARY

'Confidential information' means information whose disclosure is forbidden by statute or by order of the Court or by the terms on which it has been furnished by a Government Department.

'Exempt information' means information which fits into one of the categories listed in the Act. Of most interest to the Business is the category 'information relating to the financial or business affairs of any person other than the Authority'. The words 'financial or business affairs' include contemplated as well as past, current or now planned activities.

Appendix F
Environmental impact assessment (EIA)

SCOPE AND PURPOSE

The EEC directive on environmental impact assessment has been enacted in England and Wales in the Town and Country Planning (Assessment of Environmental Effects) Regulations 1988 and in Scotland in the Town and Country Planning (Environmental Assessment – Scotland) Regulations 1988 both of which came into force on 15 July 1988.

The regulations require developers to follow the procedure set out as a flow chart in Figure 1.1. A local authority planning department cannot now accept a planning application for any development of a type listed in Schedule 1 of the 1988 Act unless the procedure outlined in Figure 1.1 has been followed.

Although guidance is provided in the flow chart for determining whether an environmental impact outline (EIO) and/or an environmental impact assessment and statement (EIS) are required, judgement is usually necessary and so early advice should be sought.

RESPONSIBILITIES

The exact wording of an EIO and an EIS should have the agreement of the project manager and business representative and the relevant environmental and safety specialists for the site concerned.

GUIDANCE

The EIO for Schedule 2 of the 1988 Act projects will take the form of a brief for the planning officers. The information in the EIO will be put on public display. The content of the EIO should be discussed with the planning officials before submission, but will typically cover the following sections.

Background

This should provide an outline of the reasons why the business is initiating the project and its historical context.

Technology

The nature of the proposed technology and whether it is new or existing should be revealed. All benefits such as improvements in the working environment, to safety, etc should be stated.

Investment

A rough indication of the size of the financial investment may be appropriate for large projects.

Process description

An outline of the process chemistry and other detail as necessary.

Employment

Will the project bring new jobs to the local authority area?

Choice of site

The links with other units on the site should be revealed.

Chemicals handled

This should indicate the main chemical materials handled and should emphasize the owner's experience when relevant.

Gaseous emissions

Indicate if the means for control of gaseous emissions have been agreed with HMIP.

Aqueous effluents

These should normally be within the existing consent parameters. This should be stated. If this is not the case indicate why.

Other wastes

If there are other wastes arising indicate how they will be disposed of.

Noise

Normally new plants should be designed so that boundary noise levels are not measurably increased. If the development is likely to cause any problem, the statement on noise will need careful consideration.

Dust

Potential problems should be indicated.

Odour

Any special problems should be mentioned.

Traffic

The increase or decrease in road and rail traffic as a result of the proposed development should be quantified.

Visual impact

It should be made clear whether the proposal will have significant visual impact.

Index

Page numbers appearing in **bold** refer to figures and page numbers appearing in *italic* refer to tables.